Plasma Electrolytic Oxidation (PEO) Coatings

Plasma Electrolytic Oxidation (PEO) Coatings

Editors

Marta Mohedano
Beatriz Mingo

MDPI • Basel • Beijing • Wuhan • Barcelona • Belgrade • Manchester • Tokyo • Cluj • Tianjin

Editors
Marta Mohedano
Universidad Complutense de Madrid
Spain

Beatriz Mingo
The University of Manchester
UK

Editorial Office
MDPI
St. Alban-Anlage 66
4052 Basel, Switzerland

This is a reprint of articles from the Special Issue published online in the open access journal *Coatings* (ISSN 2079-6412) (available at: https://www.mdpi.com/journal/coatings/special_issues/PEO_coatings).

For citation purposes, cite each article independently as indicated on the article page online and as indicated below:

LastName, A.A.; LastName, B.B.; LastName, C.C. Article Title. *Journal Name* **Year**, *Volume Number*, Page Range.

ISBN 978-3-0365-0552-7 (Hbk)
ISBN 978-3-0365-0553-4 (PDF)

Cover image courtesy of Marta Mohedano.

© 2021 by the authors. Articles in this book are Open Access and distributed under the Creative Commons Attribution (CC BY) license, which allows users to download, copy and build upon published articles, as long as the author and publisher are properly credited, which ensures maximum dissemination and a wider impact of our publications.

The book as a whole is distributed by MDPI under the terms and conditions of the Creative Commons license CC BY-NC-ND.

Contents

About the Editors vii

Marta Mohedano and Beatriz Mingo
Special Issue: Plasma Electrolytic Oxidation (PEO) Coatings
Reprinted from: *Coatings* 2021, *11*, 111, doi:10.3390/coatings11010111 1

Frank Simchen, Maximilian Sieber, Alexander Kopp and Thomas Lampke
Introduction to Plasma Electrolytic Oxidation—An Overview of the Process and Applications
Reprinted from: *Coatings* 2020, *10*, 628, doi:10.3390/coatings10070628 3

Shun-Yi Jian, Mei-Ling Ho, Bing-Ci Shih, Yue-Jun Wang, Li-Wen Weng, Min-Wen Wang and Chun-Chieh Tseng
Evaluation of the Corrosion Resistance and Cytocompatibility of a Bioactive Micro-Arc Oxidation Coating on AZ31 Mg Alloy
Reprinted from: *Coatings* 2019, *9*, 396, doi:10.3390/coatings9060396 23

Lara Moreno, Marta Mohedano, Beatriz Mingo, Raul Arrabal and Endzhe Matykina
Degradation Behaviour of Mg0.6Ca and Mg0.6Ca2Ag Alloys with Bioactive Plasma Electrolytic Oxidation Coatings
Reprinted from: *Coatings* 2019, *9*, 383, doi:10.3390/coatings9060383 41

Xiaopeng Lu, Yan Chen, Carsten Blawert, Yan Li, Tao Zhang, Fuhui Wang, Karl Ulrich Kainer and Mikhail Zheludkevich
Influence of SiO_2 Particles on the Corrosion and Wear Resistance of Plasma Electrolytic Oxidation-Coated AM50 Mg Alloy
Reprinted from: *Coatings* 2018, *8*, 306, doi:10.3390/coatings8090306 55

Anna Kozelskaya, Gleb Dubinenko, Alexandr Vorobyev, Alexander Fedotkin, Natalia Korotchenko, Alexander Gigilev, Evgeniy Shesterikov, Yuriy Zhukov and Sergei Tverdokhlebov
Porous CaP Coatings Formed by Combination of Plasma Electrolytic Oxidation and RF-Magnetron Sputtering
Reprinted from: *Coatings* 2020, *10*, 1113, doi:10.3390/coatings10111113 67

Salih Durdu
Characterization, Bioactivity and Antibacterial Properties of Copper-Based TiO_2 Bioceramic Coatings Fabricated on Titanium
Reprinted from: *Coatings* 2019, *9*, 1, doi:10.3390/coatings9010001 79

Bernd Engelkamp, Björn Fischer and Klaus Schierbaum
Plasma Electrolytic Oxidation of Titanium in H_2SO_4–H_3PO_4 Mixtures
Reprinted from: *Coatings* 2020, *10*, 116, doi:10.3390/coatings10020116 99

Rubén del Olmo, Marta Mohedano, Beatriz Mingo, Raúl Arrabal and Endzhe Matykina
LDH Post-Treatment of Flash PEO Coatings
Reprinted from: *Coatings* 2019, *9*, 354, doi:10.3390/coatings9060354 111

Hao-Ren Lou, Dah-Shyang Tsai and Chen-Chia Chou
Correlation between Defect Density and Corrosion Parameter of Electrochemically Oxidized Aluminum
Reprinted from: *Coatings* 2020, *10*, 20, doi:10.3390/coatings10010020 127

Krisjanis Auzins, Aleksejs Zolotarjovs, Ivita Bite, Katrina Laganovska, Virginija Vitola, Krisjanis Smits and Donats Millers
Production of Phosphorescent Coatings on 6082 Aluminum Using $Sr_{0.95}Eu_{0.02}Dy_{0.03}Al_2O_{4-\delta}$ Powder and Plasma Electrolytic Oxidation
Reprinted from: *Coatings* **2019**, *9*, 865, doi:10.3390/coatings9120865 . . . . **139**

Alexander Sobolev, Tamar Peretz and Konstantin Borodianskiy
Fabrication and Characterization of Ceramic Coating on Al7075 Alloy by Plasma Electrolytic Oxidation in Molten Salt
Reprinted from: *Coatings* **2020**, *10*, 993, doi:10.3390/coatings10100993 . . . . **151**

About the Editors

Marta Mohedano completed her PhD in Materials Science and Technology (Extraordinary Award) from Complutense University of Madrid-UCM (Spain) in 2011 with the support of a UCM Pre-doctoral Fellowship, including predoctoral stays at the University of Cambridge (UK) for six months. She then continued at the UCM with a postdoctoral research contract until 2013, when she got the prestigious Humboldt Research Fellowship for Postdoctoral Researchers and moved to Helmholtz Zentrum Geesthacht—HZG (Germany). After 32 months at HZG, she reincorporated to UCM in June 2016, firstly as a Juan de la Cierva Incorporación Fellow, then in 2017 as a Principal Investigator of the national project MAT2015-73355-JIN under the umbrella of the National Programme for Research Aimed at the Challenges of Society, and since 2019, as a Ramon y Cajal Researcher. Dr. Mohedano has contributed to the field of corrosion and protection of light alloys (up to 2021) with a total of 77 publication with >2000 citations and an h-index = 29 (Scopus), more than 45 communications to national/international conferences, participation in 13 research projects, 14 technology transfer contracts and 1 utility model. In 2015, she was awarded the international EFC—Kurt Schwabe Prize, presented every three years to a young scientist below 35 years of age in recognition of his or her scientific and technical contribution to the field of corrosion of materials. She has contributed to more than 600 h of lectures at UCM and HZG (including training and supervision of more than 20 bachelor's/master's/PhD students) and has participated in scientific and organization committees (Young EFC, Local organizer Eurocorr, Advanced Chemistry Symposium), dissemination activities (Escuelab, in-school workshops) and the revision of manuscripts (Corr. Sci, JES, Surf. Coat. Tech, etc.).

Beatriz Mingo defended her PhD in 2016 at Universidad Complutense de Madrid (UCM, Spain), for which she received the Extraordinary Doctorate Award. During her PhD, she studied a range of strategies to improve the corrosion resistance of light alloys used in the automotive industry to reduce a vehicle's fuel consumption. In 2013, the British-Spanish Society granted her a scholarship to complete her academic training in 3D characterization techniques at The University of Manchester (UoM) with Prof Peter Skeldon. She also carried out two research stays at the Helmholtz Zentrum Geesthacht (Germany) in 2015 and 2016, at Prof Zheludkevich's group, where she learnt the fundamentals of active protection, with the development of coatings based on layered double hydroxides formed on Al-based composites. In 2016, she obtained the Young Scientist Grant from the European Federation of Corrosion, whose objective is to promote knowledge exchange between early career and senior scientists. In 2017, Dr Mingo was awarded a Humboldt Research Fellowship for Postdoctoral Researchers, which is a two-year German fellowship for independent researchers. In 2018, she was appointed Presidential Fellow at The University of Manchester aimed at early-career academics who can deliver world-leading research and teaching, and become the inspiring leaders of the future. This included a £20k start-up package to purchase consumables, small lab equipment and travel expenses. Moreover, the PI is also involved in the EPSRC NetworkPlus In Digitalised Surface Manufacturing, and she was recently awarded some funds (£25k) from the Network to carry out a Feasibility Study on Digitalisation of the particulate aerosol deposition process in collaboration with Dr Hall and Dr Bodjo. She was recently awarded (2021) with a EPSRC New Investigator Award to develop her project "High-Performance Smart Ceramic Coatings on Light Alloys".

Editorial
Special Issue: Plasma Electrolytic Oxidation (PEO) Coatings

Marta Mohedano [1,*] and Beatriz Mingo [2]

1 Departamento de Ingeniería Química y de Materiales, Facultad de Ciencias Químicas, Universidad Complutense, 28040 Madrid, Spain
2 Department of Materials, The University of Manchester, Oxford Road, Manchester M13 9PL, UK; beatriz.mingo@manchester.ac.uk
* Correspondence: mmohedan@ucm.es

Citation: Mohedano, M.; Mingo, B. Special Issue: Plasma Electrolytic Oxidation (PEO) Coatings. *Coatings* 2021, *11*, 111. https://doi.org/10.3390/coatings11010111

Received: 13 January 2021
Accepted: 18 January 2021
Published: 19 January 2021

Publisher's Note: MDPI stays neutral with regard to jurisdictional claims in published maps and institutional affiliations.

Copyright: © 2021 by the authors. Licensee MDPI, Basel, Switzerland. This article is an open access article distributed under the terms and conditions of the Creative Commons Attribution (CC BY) license (https://creativecommons.org/licenses/by/4.0/).

The demand of modern technological society for light structural materials (Al, Ti, Mg) emphasizes a combination of good corrosion resistance with wear properties and functionalized surfaces. Their extensive field of applications ranges from mechanical aspects and transport components to bioengineering. Regardless of the final application, improved tailored surfaces are required to prolong service life and reduce long-term costs.

Plasma Electrolytic Oxidation (PEO) is an exceptional candidate to achieve that goal: it enables a considerable improvement of the mechanical properties and corrosion resistance of light alloys, together with other characteristics such as improved biocompatibility. PEO is an environmentally friendly treatment process employed to produce relatively thick (10–100 μm) ceramic-like coatings on Mg, Al, Ti, and other valve metals, with incorporation of species from both the substrate and the electrolyte.

The coatings are formed under high voltages, exceeding those of dielectric breakdown, when short-lived discharges occur locally on the coating surface and the current density and temperature are greatly increased, facilitating the formation of phases normally associated with processing at relatively high temperatures.

The present special issue covers a wide range of information of PEO-coated light alloys for structural (Al, Mg) and biomedical applications (Ti, Mg) with 10 research papers and 1 review.

The review published by Simchen et al. [1] summarizes the main aspects of Plasma Electrolytic Oxidation Technique with special focus on the process kinetics and the influence of the process parameters on the process, and, thus, on the resulting coating properties, e.g., morphology and composition.

Regarding PEO on Mg alloys, three papers have been published covering biomedical aspects [2,3] and corrosion and wear performance [4]. It is reported by Jian et al. [2] the investigation of the corrosion resistance and cytocompatibility of a bioactive micro-arc oxidation coating on AZ31 Mg alloy including in vitro and vivo tests. A study covering new biocompatible Mg alloys–Al free is reported by Moreno et al. [3] focused on the degradation behaviour of Mg0.6Ca and Mg0.6Ca2Ag alloys with bioactive Plasma Electrolytic Oxidation coatings. In the case of applications of PEO coated Mg alloys for the transport industry, the research conducted by Lu et al. [4] covers the influence of SiO_2 particles on the corrosion and wear resistance of Plasma Electrolytic Oxidation-Coated AM50 Mg alloy.

Different aspects of PEO coated Ti alloys are analyzed in 3 works [5–7]. For instance, the study reported by Kozelskaya [5] is related to biomedical applications with special interest in the development of porous CaP coatings formed by combination of Plasma Electrolytic Oxidation and RF-Magnetron sputtering. Another research associated with biomedical features is published by Durdu [6] focused on the characterization, bioactivity and antibacterial properties of copper-based TiO_2 bioceramic coatings.

Another aspect reported by Engelkamp [7] is the influence of mixtures electrolytes H_2SO_4-H_3PO_4 on galvanostatically controlled Plasma Electrolytic Oxidation.

In the case of PEO coated Al alloys 4 works are dedicated to cover different features [8–11]. The research published by del Olmo et al. [8] investigates environmentally friendly alternatives to toxic Cr (VI)-based surface treatments for corrosion protection of Al alloys focused on multifunctional PEO-layered double hydroxides (LDH) coatings. Another work related to corrosion protection is the one published by Lou et al. [9] showing information about the correlation between defect density and corrosion parameter of electrochemically oxidized aluminum. Moving to other properties, the production of phosphorescent coatings is reported by Auzins [10] using $Sr_{0.95}Eu_{0.02}Dy_{0.03}Al_2O_{4-\delta}$ powder and Plasma Electrolytic Oxidation. In addition, Sobolev et al. [11] investigates the use of molten salt during the development of PEO coatings on AA7075 from the point of view of fabrication and characterization.

Data Availability Statement: Not applicable.

Conflicts of Interest: The authors declare no conflict of interest.

References

1. Simchen, F.; Sieber, M.; Kopp, A.; Lampke, T. Introduction to plasma electrolytic oxidation—An overview of the process and applications. *Coatings* **2020**, *10*, 628. [CrossRef]
2. Jian, S.-Y.; Ho, M.-L.; Shih, B.-C.; Wang, Y.-J.; Weng, L.-W.; Wang, M.-W.; Tseng, C.-C. Evaluation of the corrosion resistance and cytocompatibility of a bioactive micro-arc oxidation Coating on AZ31 Mg alloy. *Coatings* **2019**, *9*, 396. [CrossRef]
3. Moreno, L.; Mohedano, M.; Mingo, B.; Arrabal, R.; Matykina, E. Degradation Behaviour of Mg0.6Ca and Mg0.6Ca2Ag Alloys with Bioactive Plasma Electrolytic Oxidation Coatings. *Coatings* **2019**, *9*, 383. [CrossRef]
4. Lu, X.; Chen, Y.; Blawert, C.; Li, Y.; Zhang, T.; Wang, F.; Kainer, K.U.; Zheludkevich, M. Influence of SiO_2 particles on the corrosion and wear resistance of plasma electrolytic oxidation-coated AM50 Mg alloy. *Coatings* **2018**, *8*, 306. [CrossRef]
5. Kozelskaya, A.; Dubinenko, G.; Vorobyev, A.; Fedotkin, A.; Korotchenko, N.; Gigilev, A.; Shesterikov, E.; Zhukov, Y.; Tverdokhlebov, S. Porous cap coatings formed by combination of plasma electrolytic oxidation and RF-magnetron sputtering. *Coatings* **2020**, *10*, 1113. [CrossRef]
6. Durdu, S. Characterization, bioactivity and antibacterial properties of copper-based TiO_2 bioceramic coatings fabricated on titanium. *Coatings* **2019**, *9*, 1. [CrossRef]
7. Engelkamp, B.; Fischer, B.; Schierbaum, K. Plasma Electrolytic Oxidation of Titanium in H_2SO_4–H_3PO_4 Mixtures. *Coatings* **2020**, *10*, 116. [CrossRef]
8. Del Olmo, R.; Mohedano, M.; Mingo, B.; Arrabal, R.; Matykina, E. LDH Post-Treatment of Flash PEO Coatings. *Coatings* **2019**, *9*, 354. [CrossRef]
9. Lou, H.-R.; Tsai, D.-S.; Chou, C.-C. Correlation between Defect Density and Corrosion Parameter of Electrochemically Oxidized Aluminum. *Coatings* **2020**, *10*, 20. [CrossRef]
10. Auzins, K.; Zolotarjovs, A.; Bite, I.; Laganovska, K.; Vitola, V.; Smits, K.; Millers, D. Production of Phosphorescent Coatings on 6082 Aluminum Using $Sr_{0.95}Eu_{0.02}Dy_{0.03}Al_2O_{4-\delta}$ Powder and Plasma Electrolytic Oxidation. *Coatings* **2019**, *9*, 865. [CrossRef]
11. Sobolev, A.; Peretz, T.; Borodianskiy, K. Fabrication and Characterization of Ceramic Coating on Al7075 Alloy by Plasma Electrolytic Oxidation in Molten Salt. *Coatings* **2020**, *10*, 993. [CrossRef]

Review

Introduction to Plasma Electrolytic Oxidation—An Overview of the Process and Applications

Frank Simchen [1],*, Maximilian Sieber [2], Alexander Kopp [3] and Thomas Lampke [1]

[1] Materials and Surface Engineering Group, Institute of Materials Science and Engineering, Chemnitz University of Technology, 09107 Chemnitz, Germany; thomas.lampke@mb.tu-chemnitz.de
[2] Department of Corrosion Protection and Testing, EXCOR Korrosionsforschung GmbH, Magdeburger Strae 58, 01067 Dresden, Germany; maximilian.sieber@excor.de
[3] Meotec GmbH, Triwo Technopark Aachen, Philipsstraße 8, 52001 Aachen, Germany; Alexander.Kopp@meotec.eu
* Correspondence: Frank.Simchen@mb.tu-chemnitz.de; Tel.: +49-371-531-30115

Received: 28 April 2020; Accepted: 28 June 2020; Published: 30 June 2020

Abstract: Plasma electrolytic oxidation (PEO), also called micro-arc oxidation (MAO), is an innovative method in producing oxide-ceramic coatings on metals, such as aluminum, titanium, magnesium, zirconium, etc. The process is characterized by discharges, which develop in a strong electric field, in a system consisting of the substrate, the oxide layer, a gas envelope, and the electrolyte. The electric breakdown in this system establishes a plasma state, in which, under anodic polarization, the substrate material is locally converted to a compound consisting of the substrate material itself (including alloying elements) and oxygen in addition to the electrolyte components. The review presents the process kinetics according to the existing models of the discharge phenomena, as well as the influence of the process parameters on the process, and thus, on the resulting coating properties, e.g., morphology and composition.

Keywords: plasma electrolytic oxidation; surface treatment; corrosion; wear; medical engineering; aluminum; magnesium; titanium

1. Introduction

Plasma electrolytic discharge phenomena were first described by Sluginov around 1880 [1]. In the 1920s, these were systematically examined by Güntherschulze and Betz as an aspect of the development of electrolytic capacitors [2]. In the early 1970s, Brown and co-workers derived a method from the phenomena described to produce ceramic conversion layers on Al substrates in alkaline electrolytes, which they referred to as Anodic Spark Deposition (ASD) [3]. In the 1980s and 1990s, the working groups of Snezhko, Markov, Kurze and others made further progress, which led to the first practical applications [4–6].

Since then, the technological and commercial introduction of the PEO into practice by specialized companies has succeeded: Keronite (GB), Meotec, Innovent, AaST, Cermanod (DE), Hirtenberge (AT), Tekniker (ES) IBC (US), Manel (RU) MAO Environmental Production Technology (CN), the related specialist literature began to split up thematically. To respect the immense research activity in the field of PEO, current reviews are increasingly dealing with key topics such as special substrate materials [7–9], particle incorporation [10,11], selected technological properties [12,13] and characteristics of the discharge phenomena [14,15]. While excellent reviews from past decades [16,17] on the basics of the PEO exist, their coverage of current developments is limited.

The present work should, therefore, provide an introduction to the topic and convey both, the fundamental basics and the approaches of modern development trends and form a link to current fields of application.

2. Principles of Plasma Electrolytic Oxidation

Plasma electrolytic oxidation (PEO) is also referred to as micro-arc oxidation (MAO), anodic spark deposition (ASD), plasma chemical oxidation (PCO), or anodic oxidation by spark discharge (ANOF, German: anodische Oxidation unter Funkenentladung). It is a conversion coating process for the surface refinement of several metallic materials, which tend to passivity in adequate aqueous electrolytes. In the first decade of the 21st century, PEO had also been developed for iron-based materials, which exhibit usually a poor passivation behavior. Table 1 shows a brief overview about literature known PEO processes, categorized according to application and substrate material.

Table 1. Selected applications and examples for the PEO of different materials.

Application/Motivation	Substrate
Corrosion and wear protection	Al [18,19], Mg [20,21], Ti [22,23], Zn [8], brass [24] Fe/Steel [25–28], Nb [29,30], Be [31], Ta [7,32] c-graphite materials [9]
adjustment of radiation behavior improved thermal emission, lowered absorbance lowered optical reflection	Ti [33], Mg [34] Ti, Mg [35]
decorative puropse (by coloring)	Al [36,37], Mg [38]
Improvement of thermal isolation	Al [39]
Medical Issues Formation of hydroxyapatite (HA) for improved bioactivity Adjustment of the degradation in contact with body tissue	Zr [40], Ta [41], Mg [42], Ti [43,44] Mg [42]
only selective scientific description photoluminescence Catalytic activity	Hf [45,46] Fe [47,48]

The process is characterized by discharge, which develop under a strong electric field in a system consisting of the substrate, the oxide layer, a gas envelope, and the electrolyte, and it specifically determines the morphology, as well as the composition of the produced coatings. The electric breakdown in this system establishes a plasma state, in which, under anodic polarization, the substrate material is converted to a compound, comprised of the substrate material itself (including alloying elements), oxygen, and the electrolyte components.

The PEO process originates in the anodic oxidation of metals. When a metal electrode is polarized anodically in an electrolyte, different reactions are possible. A metal electrode, which is insoluble in the electrolyte, will lead to the evolution of oxygen (water electrolysis). If the metal electrode is soluble in the electrolyte, salts comprised of the electrode material and electrolyte components will occur, and the electrode will be consumed. The third possible reaction is the reaction of the anode material with the oxygen provided from the electrolyte to form a thin passive film, which itself is not or barely soluble in the electrolyte. Passive films are usually composed of oxides or hydroxides of the anode material, but more complex compounds of substrate and electrolyte components are also known to be formed. In order to prevent the reaction layer from flaking off, the unit cell volume of the reaction products must be in a favorable ratio to the volume of the unit cells of the substrate material. In case of metal oxides, this is characterized by the so-called Pilling-Bedworth ratio (PBR). For hydroxides and more complex compounds, the relationship is described as product/metal ratio PMR. [49,50].

It is crucial for a technologically-relevant passive film formation that it does not exhibit electron conductivity, but rather ion conductivity [2]. This behavior is strongly dependent on the combination of electrode metal and electrolyte. Figure 1 summarizes the possible current density-potential behavior of an anodically polarized electrode in an electrolyte.

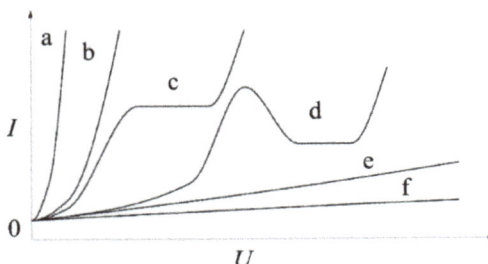

Figure 1. Principal types of the current density-potential behavior of an anodically polarized electrode in an electrolyte in accordance to Kurze [5]: a/b - dissolution, c - passivation in small potential range, d - complex behavior, e/f - passivation.

Only the passivating and, with some limitations, the complex behavior with a passive and a transpassive region are suitable for formation of reaction layers, which are appropriate for PEO initiation. A general overview over the chemical reactions that proceed during the growth of oxide e.g., hydroxide is given in Table 2.

Table 2. Generalized chemical reactions proceeding during oxide or hydroxide layer formation.

Reaction			Description	Location
H_2O	↔	$2H^+ + O^{2-}$	water dissociation	-
Me	→	$Me^{n+} + ne^-$	metal oxidation/hydration	anode
$xMe^{n+} + yO^{2-}$	→	Me_xO_y		
$xMe^{n+} + y(OH)^-$	→	$Me_x(OH)_y$		
$nH^+ + ne^-$	→	$0.5nH_2$	hydrogen evolution	cathode
$xMe + yH_2O$	→	$Me_xO_y + yH_2$	overall reaction	-

The formation of the ion-conductive oxide layer results in significant electric resistance. In a current-controlled, galvanostatic process, it is characterized by a steep increase in the cell voltage within the first few seconds. Increasing electric field strengths are necessary to realize a further current flow and further oxide growth.

The rise in the anodic potential over the electrolyte/oxide/electrode system leads to the partial formation of a gas film around the electrode. This film consists of oxygen, arising from the electrochemical, or in later process states with high local energy input, thermal decomposition or vaporization of water. Additionally, the formed film further increases the electric resistance in the system electrolyte/gas/oxide/electrode. Thus, perpetuation of the current flow requires an increase of the potential until the strength of the electric field in the aforementioned system reaches a critical value, and the breakdown occurs. All this typically happens within the first minute of the process [17].

The breakdown of the system is mainly affected by the substrate material and the electrolyte composition, while it is independent of the current density, temperature, surface roughness, electrolyte movement, and the history of the system [51,52].

By injection of electrons at the electrolyte/gas interface, which acts as a quasi-cathode (equipotential area of the electric field), a discharge channel evolves and penetrates the oxide layer. Within the discharge channel, thermally-activated ions originating from the substrate metal are ejected and move away from the substrate, due to the migration in the electric field, while oxygen ions move towards the substrate. The oxide is then formed in a reaction of the substrate ions and oxygen ions and is deposited in the boundary regions of the channel. The discharge channel is characterized by a local current flow of an order of magnitude several kiloamperes per square centimeter under a high electric field, which results in enormous energy density. Thus, temperatures of several thousand Kelvin may occur locally. At the beginning of the discharge event, the breakdowns are concentrated on those surface regions with the highest electric field strength. Small, discrete micro-discharges are visible. In a galvanostatic process, the steep rise of the cell voltage gives way to a substantially less-pronounced increase or even slight decrease of the voltage. During the evolution of the process, the discharges become larger, while fewer discrete breakdown events are visible. Therefore, the energy within the discharges increases. This leads to a series of consequences: (1) The formation of high-temperature crystalline phases is promoted (e.g., α-alumina). This does not only imply the direct formation of these phases, but also the phase transformation of already-formed oxide in later process stages. (2) The direct vicinity of the discharge channel is heated. Since the breakdown voltage of the electrolyte/gas/oxide/electrode system decreases with increasing temperature, the initiation of a new discharge in the vicinity of a former discharge is promoted. However, no negative effect on the electrode metal occurs, since the thermal influences of the discharge events are limited to small volumes. The substrate usually does not suffer significant heating. (3) Re-melting of the oxide occurs. Since the dissipation of heat towards the electrolyte is generally higher than that towards the substrate, near-substrate regions of the coatings can be rather loose morphology in this region. (4) Large discharges can destroy the formed oxide coating. The occurrence of such detrimental discharges is dependent on the process parameters. It can take place several minutes to hours after the initiation of the first discharges and should be avoided. In a galvanostatic process, the cell voltage generally continues to increase at a relatively low rate, but it usually drops instantaneously with the occurrence of large and deteriorating discharge. The time on which this stage of the process is reached is strongly dependent on the process parameters. It is advisable to choose process time and parameters so that this critical stage is avoided [53,54].

An example of the discharges evolution on a sample of the magnesium alloy AZ31 during the PEO in an alkaline silicate electrolyte is given in Figure 2. The brighter regions reflect the discharge action with brightness as a measure for the discharge intensity. The characteristic growth of discharges and the decrease of their number are observable in Figure 2 as well.

Figure 2. Evolution of discharge distribution and intensity in relation to the process time for the PEO of an AZ31 magnesium alloy in an alkaline silicate electrolyte. Brightness correlates with discharge intensity on the respective surface region of the sample.

The three-layered structure of PEO coatings which typically occurs is shown schematically in Figure 10. The oxide coating is comprised of a nanometer-thin and, according to the described model, nearly defect-free (amorphous) barrier layer at the oxide/substrate interface [55,56], a rather compact working layer, and a loose outer layer, the so-called technological layer [18,19]. The thickness of the compact and the technological layer can stretch from a few microns to several hundred microns, depending on the conditions under which the PEO process is performed. Hence, the entire process is controlled by the electrical field and takes place under high electric potentials, and the PEO shows a very good throwing power which results in a homogenous layer thickness distribution, even on working pieces with complex geometry.

In the following, the different dependencies of the PEO process, primarily the substrate composition, the electrolyte used, the applied electrical regime, as well as their interaction with each other are discussed. Based on this, the formation and the technological properties of the resulting PEO layers are discussed and selected current application options are presented. Figure 3 summarizes the order of the focal points in this review using a schematic representation of the process steps during the plasma electrolytic oxidation.

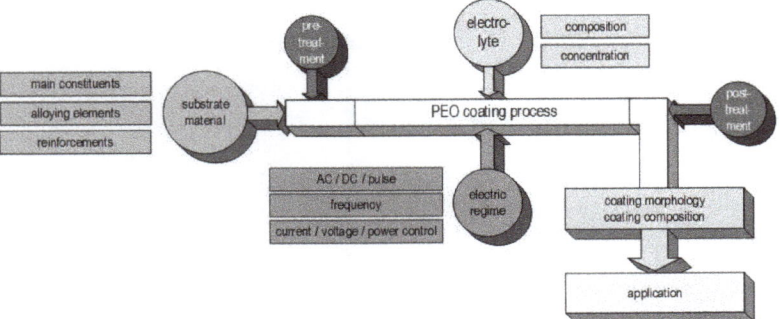

Figure 3. Schematic graphic representation of the PEO process and its dependencies from the uncoated substrate over the process parameters to the application.

3. Process Parameters and Coating Properties

3.1. Substrate

Plasma electrolytic oxidation in common low concentration aqueous electrolytes is at its core a conversion coating process. Hence, the nature of the oxide strongly depends on the substrate composition. The metal ions participating in the electrochemical reactions during the PEO process (Table 1) are determined by the treated material. Generally, oxides of the substrate metal are the main constituents of the coatings. The substrate conversion naturally includes alloying elements and precipitates in the metal as well as reinforcement phases in case of metal matrix composites. For aluminum, this is shown in Figures 4 and 5. Figure 4 shows a scanning electron microscopy (SEM) picture of the oxide/substrate interface in the cross section of an oxide coating, produced by PEO in an alkaline silicate electrolyte on thermally-sprayed aluminum comprised of copper particles in BSE-mode (element contrast).

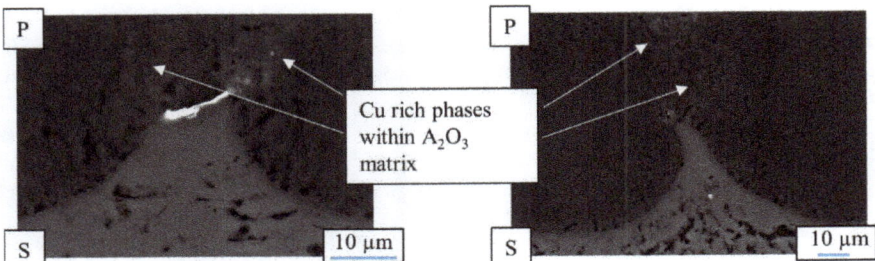

Figure 4. Cross section of the oxide/substrate interface of a PEO coating produced on thermally sprayed aluminum comprised of copper particles (P – PEO layer, S – Substrate). The brighter regions reflect a higher atomic number of the displayed material (SEM in BSE-mode – element contrast).

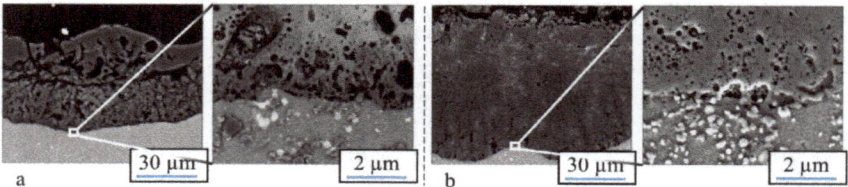

Figure 5. Cross section of the oxide produced in a PEO process on a SiC- (**a**) and a Al$_2$O$_3$ - (**b**) particle-reinforced aluminum alloy of the 2000 series [57]. (Reprinted from [57]. Copyright from 2016 IOP).

The bright fraction in the image represents a copper particle in the substrate. In the oxide above, the copper is obviously incorporated. It is also noticeable that the copper particle obstructs the conversion of the substrate, as in the surroundings, more of the substrate material has been consumed by the conversion coating process. Figure 5 shows a cross section of an oxide coating produced on an aluminum matrix composite (AMC) reinforced with silicon carbide or aluminum oxide particles.

It is obvious that the presence of SiC in the aluminum matrix leads to an increased number of flaws in the oxide, which are correlated to gas evolution at the electrically conductive particles. Remains of the particles, which are also partly converted, can be found in the pores of the coatings. However, under the high over-potential, the conversion of the silicon carbide particles takes place, presumably under gas evolution. This also results in a lower coating thickness, since the current efficiency is deteriorated by the evolution of gas. Unlike the presence of SiC, the presence of electrically non-conductive Al$_2$O$_3$ particles within the substrate does not affect the morphology or mechanical properties of the resulting PEO layer [57].

For aluminum in general, the existence of alloying elements like zinc, copper, or magnesium is likely to impede the transition from the metastable γ-phase to the high-temperature α-phase [58,59]. In addition, the alloying elements significantly influence the oxide phase distribution within the PEO layer [18]. Meanwhile, for magnesium alloys, the achievable coating thickness grows with an increasing content of aluminum or rare earths [55].

However, the coating composition and morphology are, not only influenced by the substrate material, but can also be altered by the incorporation of electrolyte constituents, as will be shown in the following section.

3.2. Electrolyte

Section 2 includes a classification of electrolyte components with regard to their passivating or dissolving behavior towards the substrate. Nevertheless, other classifications are possible. With regard to the incorporation of foreign compounds into the oxide coating, electrolytes are classified as follows: (1) electrolytes leading only to oxygen incorporation, (2) electrolytes leading to the incorporation of foreign compounds by anions, (3) electrolytes leading to the incorporation of foreign compounds by cations, and (4) electrolytes containing macroscopic particles, which are incorporated into the oxide by cataphoretic processes [17].

Common salts for the PEO of aluminum, magnesium, titanium, and their alloys in alkaline media are, amongst others, silicates, phosphates, aluminates, fluorides, borates, and stannates. Especially for the PEO of magnesium and its alloys, acidic or pH-neutral electrolyte compositions are used instead of alkaline electrolytes, e.g., fluoric acid, phosphoric acid, and/or boric acid in combination with organic additives, c.f. [60,61]. By incorporating elements provided by the electrolyte, the composition of the oxide can be altered substantially.

Some examples are shown in Figure 6, which contains x-ray diffractograms of uncoated Mg-AZ31 (a) as well as samples of the same material, which were treated by PEO within different alkaline media (b–d). The use of a low concentration phosphate electrolyte leads to the formation of a magnesia layer by substrate conversion. On the other side, highly concentrated solutions allow for the incorporation of electrolyte constituents into the resulting layer. In this way, PEO layers are formed in aluminate-rich solutions, which contain high proportions of crystalline $MgAl_2O_4$ spinels (6c) [21]. By using silicate-rich electrolytes, it is even possible to completely replace MgO in the PEO layer and to produce coatings, consisting of amorphous Mg / Si mixed oxides (6d), e.g., [20]. Since aluminum- or silicon-containing mixed oxides are usually superior over magnesia or titania in regards to their hardness and chemical resistivity, this approach is relevant for the PEO of magnesium and titanium alloys. Therefore, the method is converted from a conversion process to a mixed form of conversion and deposition. Another prominent example of the incorporation of electrolyte constituents into the oxide coatings is the formation of hydroxyapatite-containing PEO coatings on titanium from calcium- and phosphorus-containing electrolytes, c.f. [43,44]. Additionally, the coloring of working pieces by PEO for decorative or optical applications (e.g., blackened by use of vanadate ions) is already in practical use, c.f. [34,36].

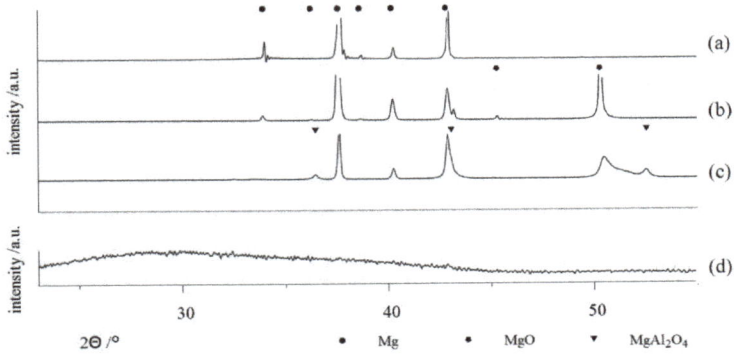

Figure 6. X-ray Diffraction (XRD)-diffractograms on bare AZ31 substrate, (a) and PEO coatings on AZ31 generated within several alkaline electrolytes; a low concentration phosphate solution, designed for substrate conversion, (b) high concentration silicate and aluminate electrolytes, designed for focused incorporation of electrolyte constituents (c,d) [20,21] (Apapted from [20,21]).

For the PEO of aluminum, the widely-used silicate components lead to the incorporation of large amounts of silicon oxides or alumina-silica-mixed-phases (e.g., mullite), have a passivating effect on the aluminum substrate, while also not dissolving the formed alumina. This generally increases the achievable coating thickness and results in a compact morphology of the coatings. In contrast, the dissolution of alumina in strongly alkaline solutions [62,63] allow for an adjustment of the coating growth through the addition of hydroxide to the electrolyte [4]. However, recent investigations indicate that this mechanism mainly affects the amorphous alumina phases, c.f. [18].

3.3. Electric Regime

The electric regime during PEO can be determined by the control parameter (current density or cell voltage), the type of the supplied parameter (direct, alternating, pulse current/voltage), and the definition of the regime (frequency, breaks, limits etc.). Under direct current or voltage, the discharge events become ever more intense during the progression of the process. This includes large discharges with long life periods, which can have a deteriorating effect on both, the formed oxide and the substrate and thus lead to irreparable defects. This behavior results from excessive energy transfer and heat release. Therefore, pulse or alternating current or voltage regimes are used to limit the effect of the strong discharges and to facilitate the formation of thick oxide coatings up to a few hundred microns. Thus, the PEO process can be prolonged, the number of defects in the coatings decreased, and the formation of a thick technological layer on top of the coating is impeded. In the following, the effects of a pulsed regime and of an alternating regime shall be discussed.

In a pulse regime, the control parameter (current or voltage) is regularly set to a low value, typically to zero. During the formation of the barrier layer at the beginning of the process (the pre-spark stage), this only affects the process substantially, if the substrate material or the oxide are dissolved by the electrolyte.

The break in current flow regime gives time for chemical dissolution and can lead to a slower growth of the barrier oxide. In most cases, the barrier layer growth accelerates, since the break allows for the compensation of potential concentration gradients. Furthermore, the repeated steep rise of the control parameter results in a significant mass/charge transfer towards the substrate. After formation of the barrier layer, the system behaves like a capacitor. This includes that current flow occurs only above of a certain potential threshold. At each rising edge of the pulse, collision ionization is likely to occur and implies that the initiation of short, but intense discharge events are favored, as opposed to those favored in a direct current or voltage regime. At the falling edge of the pulse, the current flow is interrupted, and thus the length of the discharge periods are limited. Using this strategy, the PEO process can be performed for a longer duration without the development of large and deteriorating discharges. The utilization of an alternating regime likewise limits the life period of discharges. In addition, the cathodic half period of an alternating regime can lead to a partial electrochemical reduction of the oxide. Thus, the formation of a barrier layer sufficient for discharge initiation at the beginning of the PEO process will be prolonged. During the PEO process, discharge can also occur in the cathodic branch of the electric regime. This discharge is usually less intense than the discharge in the anodic branch. A lower amount of energy is introduced into the coating. However, the cathodic discharge is also prone to heat up the oxide locally, and are thus, likely to result in a reduction of the field strength necessary for the breakdown in the following anodic branch [54].

In general, the interactions of the substrate/electrolyte-combination with the electrical regime are complex and still a subject of research. A promising approach to gain experimental access to the relationships is the analysis of the electrical process data. Since, in contrast to other electrolytic surface treatment methods, PEO results in the formation of high-ohmic layers, these affect, above all in the case of current-controlled regimes, to what extent the pre-defined electrical pulse is mapped correctly in the experimental setup. Figure 7 shows a schematic representation of these relationships. Figure 8 illustrates them using the example of PEO of Mg-AZ31 samples with identical experimental parameters using different electrolytes.

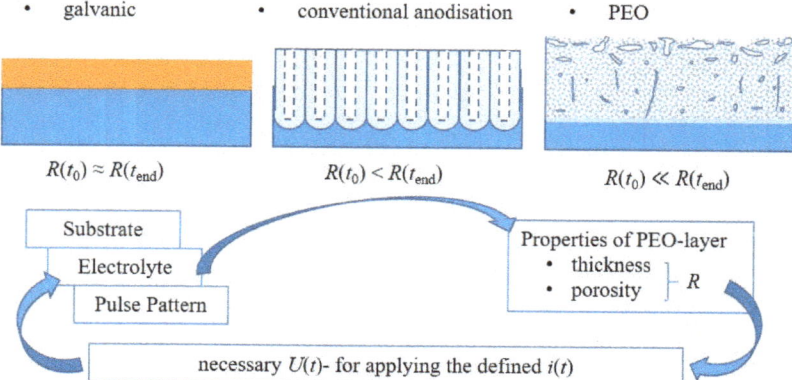

Figure 7. Schematic representation (not to scale) of obtained-layer morphology and development of layer resistance R for different electrolytic coating methods relation to the process time t, interactions of the process variables voltage $U(t)$ and current $i(t)$ pattern with the layer characteristics during the PEO.

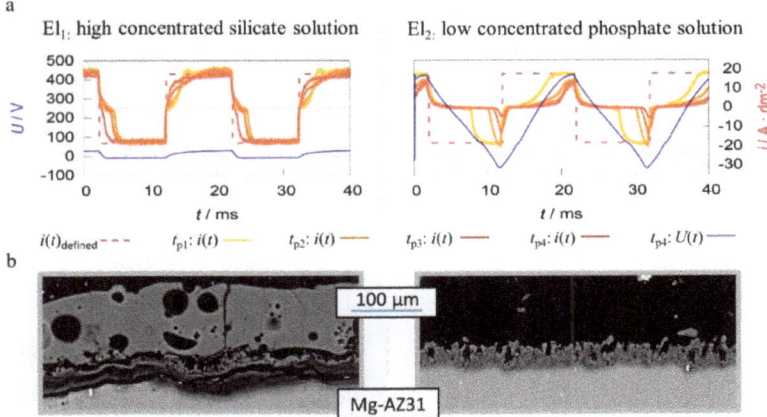

Figure 8. Pulse shape development of current-controlled PEO experiments for different process times (t_{p1-4} = 1, 5, 10 and 20 min) under identical experimental conditions in two different alkaline electrolytes $El_{1/2}$ (**a**) SEM-micrographs of the resulting layers after 22:30 min of treatment time (**b**).

The working media used are designed for focused incorporation of electrolyte constituents into the coating (El_1) and PEO layer formation by substrate conversion (El_2). The current time characteristics measured during the process within the high concentration silicate solution show that the predefined symmetric rectangular current pattern is transmitted largely correctly to the system under investigation. Furthermore, the remaining deviations between predefined and measured pulses are decreasing with progression of the process time. In contrast to the effects described above, the experiment within the low concentration phosphate electrolyte shows a significant delay in the current flow after polarity reversal in both, the cathodic and anodic partial period. The delays even increase during the process. Additionally, the experiments in electrolyte 2 result in much higher voltages over the entire treatment time. This is exemplified by the voltage pattern after 20 min and leads to a higher consumption of electrical energy.

Therefore, one might assume that the experiment in electrolyte 1 had the more desirable process characteristics. However, consideration of the micrographs displayed in Figure 8b show that, despite a significantly increased layer thickness, an insufficient layer adhesion was achieved within electrolyte 1. Conversely, the coating formed in electrolyte 2 shows a good substrate bonding. Thus, the more pronounced delay of the current flow with increasing process time is not an error in the control of the behavior of the rectifier but belongs to the PEO process characteristics. A good adhesive layer of ceramic leads to an increasing electrical resistance at the substrate/electrolyte interface, which indicates that the current flow only starts after the rectifier has readjusted to higher process voltages. If the layers produced adhere poorly or worse with increasing treatment time, this mechanism is suppressed.

This behavior is mentioned at this point in discussion because, in addition to the layer state, the ignition voltage for discharge initiation during the PEO is also influenced by the electrolyte composition. However, the electrolyte resistance (or electrolyte conductivity) is negligible in most cases [51]. The depiction of the electrical process variables in relation to each other allows further conclusions to be drawn about the underlying process characteristics. Figure 9 represents the process data shown in Figure 8 as voltage current curves (VCC), as well as the course of specific voltage values over the treatment time.

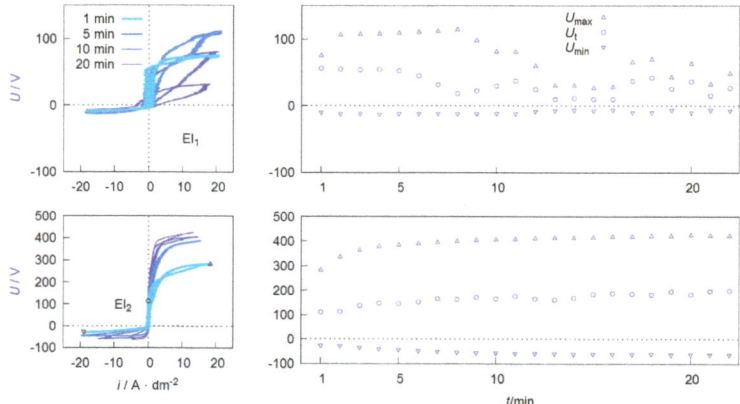

Figure 9. Voltage current curves of VCC (left) and course of maximum U_{max}, threshold U_t and minimum U_{min} voltage during the process (right) of the PEO experiments described in Figure 8.

Depending on the form of the representation, these curves are also referred to as current-voltage curves or characteristics or more generally as oscillograms. The value of the maximum U_{max} and minimum U_{min} voltage during one pulse cycle can be easily extracted from such a diagram. Furthermore it is possible to see that the downward branch of the anodic period shows a characteristic point at which the current flow stops while the voltage is still above the coordinate origin. This behavior can be observed at the process data of PEO on aluminum under various conditions, shown in several publications [64–68]. Within studies by Suminow et al. the value is called threshold voltage [1,2]. Duan et al. describes the polarization level below, when the charge carrier flow comes to a standstill as a critical conductive voltage [69]. Both research groups investigate the course of this voltage in dependence of the treatment time and assume that the value is proportional to the electrical isolation properties of the PEO layer after collapse of the plasma electrolytic discharges. Therefore, an increasing U_t indicates layer thickness growth and/or a decrease in the defect density within the layer. Insofar that this is correct, the evaluation of U_t allows, with respect to the anti-corrosive properties of the resulting PEO layers, the determination of optimal treatment times. In addition, the workload for materialographic examinations are reduced, since U_t provides automatically accessible information about the layer morphology. In contrast to microstructure images, these relate not only to a selected two-dimensional area but to the PEO layer along the entire substrate geometry. Hence, the method would be interesting also for non-destructive quality control.

Another interesting peculiarity of the VCCs shown in Figure 9 is that the ascending and descending anodic branches are not congruent. This is in accordance with recent studies by Ragov et al., which deal in detail with the analysis of electrical process data of the PEO of aluminum. Accordingly, the so-called hysteresis effects only occur in pulse regimes with cathodic components. The different conductivities in the rising and falling region of the anodic partial pulse are attributed to cathode-induced changes (CIC). These among others are substantiated by the incorporation of cations in a thin reaction layer at the substrate layer/interface of the so-called active zone. The subsequent release of the cations at the beginning of the anodic pulse leads to a brief increase in the electrical conductivity of the system. It is assumed that the few nm thick active zone has a strongly disproportionate contribution to the total electrical resistance of the layer system [70–72].

This would limit the significance of the threshold voltage mentioned above. Provided that the results can be transferred to the PEO of magnesium, the more pronounced hysteresis effects in Figure 9 for electrolyte 1 can be explained by the fact that there are significantly more cations in the electrolyte and the cathodic partial pulses are mapped almost completely.

4. Protective Properties of PEO Coatings

An example of the morphology of a PEO coating on an aluminum alloy (AlMgSi1), produced in an alkaline silicate electrolyte under the usage of a rectangular bipolar-pulsed current, is displayed in Figure 10.

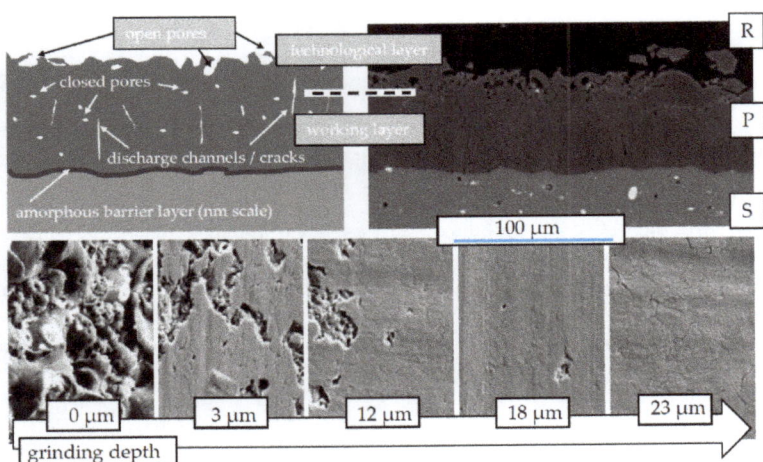

Figure 10. Morphology of a PEO coating produced on AlMgSi1 alloy, schematic representation and micrographs of a cross section (R – Embedding Resin, P – PEO-Layer, S – Substrate), as well as several top views with increasing grinding depth, which show an increasingly compact layer structure [19]. (Reprinted from [19]. Copyright from 2014 Elsevier).

In the cross section of the coating, the rough and less compact technological layer at the top of the coatings, as well as the working layer, which is characterized by numerous micro-cracks and only small flaws, are visible. The top view of the successively grinded coating exhibits the typical surface morphology of PEO coatings, in which discharge channels are visible. With successive grinding, however, the number of visible flaws and defects is substantially reduced. The coating consists of aluminum oxide in different modifications (α-, γ-, δ- and amorphous) and exhibits a hardness of approx. 9.5 GPa on the Martens scale (HM0.05/30/30). As depicted in Figure 11a, the morphology is reflected by the wear behavior (rubber wheel test according to ASTM G65), in which a significant initial wear of the coating is registered, while the subsequent increase of the sliding distance leads only to a moderate increase of the mass loss of the samples. In order to optimize hte PEO layers for tribological applications, the technological layer is sometimes removed by an additional polishing step. For classification, Figure 11b shows the appropriate volume wear of the mentioned PEO coating in comparison to other state-of-the-art wear-resistant coatings.

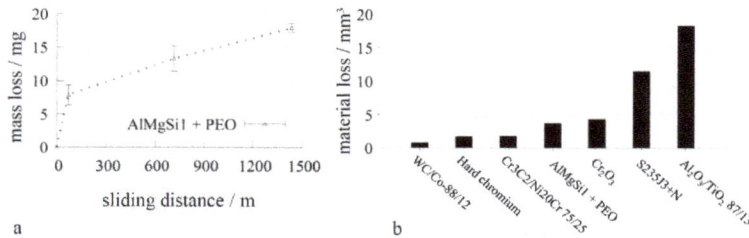

Figure 11. Wear performance of PEO coating on AlMgSi1 in the rubber wheel test: mass loss evolution in relation to the sliding distance (**a**), material loss after the testing in comparison with other coatings (**b**).

As can be seen, the PEO coating offers a wear resistance similar to materials, which are either much heavier, or the production or handling of which is unsanitary and/or harmful to the environment. Furthermore, the use of these alternative wear protection systems almost exclusively contains heavy materials with a high density. Additionally, the PEO coating provides reasonable corrosion protection. Figure 12 shows the impedance behavior of the PEO-coated AlMgSi1 alloy in comparison to the bare AlMgSi1 alloy in dilute, acidic sodium chloride solution.

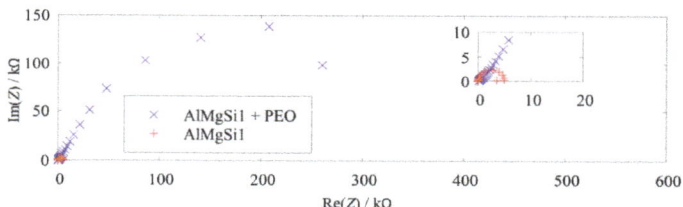

Figure 12. Result of electrochemical impedance spectroscopy of the PEO coating shown in Figure 10 in acidic dilute NaCl solution: The characteristic semicircle in the Nyquist-Plot shows that the resistance against charge transfer (= corrosion reaction) at the electrolyte/substrate interface is increased by the PEO coating.

The diameter of the characteristic semi-circle represents the resistance against charge-transfer between electrolyte and substrate. Hence, it is interpreted as an abstract measure for corrosion resistivity. For the PEO-coated surface, the charge transfer resistance is approx. 300 kΩ, while it equals 5 kΩ for a bare aluminum surface. Therefore, the resistance against the flow of current, which is correlated to corrosive attack, is increased by two orders of magnitude.

PEO coatings on magnesium and titanium materials exhibit a lower mechanical stability than those produced on aluminum alloys. This behavior is observed as the oxide phases obtained by substrate conversion like periclase (cubic MgO) or rutil and anastas (tetragonal TiO_2 modifications) have a lower hardness than the aluminum oxide modifications. For this reason, highly-concentrated process media are often used for the PEO treatment of these materials in order to shift the phase composition of the layers produced in favor of more resistant compounds through the focused incorporation of electrolyte components (see Section 3.2). Furthermore, subprocesses, which are not yet entirely understood, during the layer formation lead to PEO coatings on Mg and Ti alloys, revealing a far less compact structure than shown in Figure 10 for Al materials. In the case of magnesium, toxic fluoride compounds are known as suitable electrolyte components to improve the morphology of the resulting layers [73,74]. In general, sealing by organic or inorganic polymers can be applied as a post treatment to fill open porous cavities [75,76]. Thus, a substrate attack by corrosive media can be limited, while the structure of the coating is mechanically supported.

5. Further Functional Characteristics and Applications

Based on its low density and high wear and corrosion resistivity, PEO-coated aluminum is widely used for quickly moving parts which face aggressive atmospheres. Figure 13 shows the rotor of a turbo molecular pump with a diameter of approximately 15 cm (a) and a centering ring (b). The components are used to transport gases and for plasma etching processes, and they are working at rotation speeds of some 10,000 min^{-1} [77].

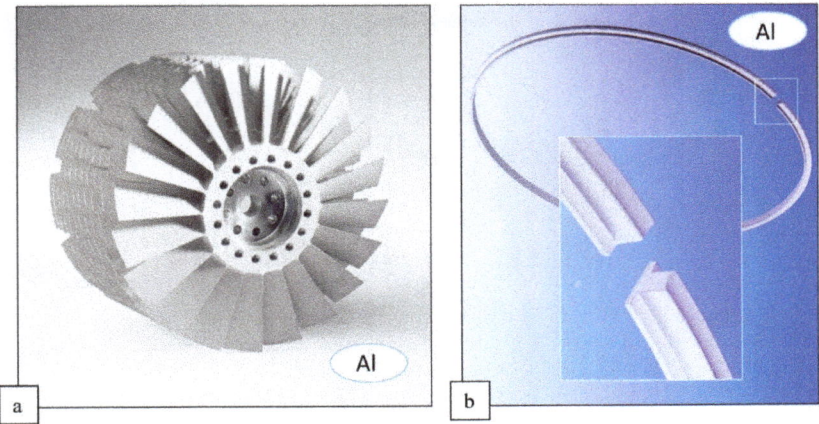

Figure 13. Application examples for PEO coatings (commercial name KEPLA-COAT®®) or Al-substrates: Rotor (a) and centering ring, (b) of turbomolecular pumps, coatings from Aalberts Surface Treatment GmbH [77].

Furthermore, PEO-coated devices are interesting for space engineering applications: The layers have excellent thermocycling resistance, due to their good substrate binding. Therefore, they withstand tremendous temperature fluctuations, which occur in space as a result of changing irradiation and shading, without the flaking of the PEO layers. An example of this is shown in Figure 14a. The photo shows a close-up of the EXPOSE experiment on the international space station ISS.

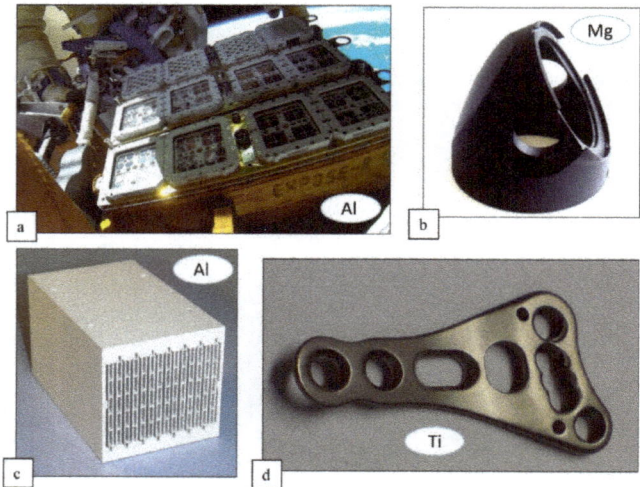

Figure 14. Application examples for PEO coatings (commercial name PCO®®) on different substrates: setup for astrobiological experiments during the *expose* program at the ISS (**a**), a heat sink (**b**), optical die-cast component (**c**), and a plate for osteosynthesis (**d**), products of INNOVENT e.V. [35].

The experimental setup is in use for astrobiological experiments. The covers are coated with PEO layers, which provide the necessary corrosion protection and, through their very low tendency towards outgassing, additionally prevent contamination of the experimental atmosphere. Low outgassing is also important to prevent the degradation of optical devices. The die-cast component depicted in Figure 14b was additionally blackened during the PEO by use of adapted electrolytes. Therefore, the light absorption is improved and the stray light effects by reflection are reduced to a minimum. Additionally, such surfaces show a convenient ratio of solar absorptance and emittance close to one, which is desirable for passive thermal control in space. Figure 14c shows a plasma-electrolytically-oxidized heat sink. The excellent throwing power of the PEO allows a uniform inner coating of the filigree channels [35].

The commercial name PCO®® stands for plasma chemical oxidation. Its different modifications shown in the Figure 14a–d have the suffixes 13-white, 12-black, 13-white and 22-bio, which represent the atomic number of the substrate dominating element and functionality.

Another field of application of PEO is medical engineering, such as the production of osseo-integrative coatings for dental and orthopedic implants. Implant materials have to meet certain requirements like biocompatibility (hemocopatibility, cytocompatibility), non-toxicity, chemical stability, and corrosion resistance. With a moderate density and a good specific strength, titanium is a material of choice for implants. Furthermore, titanium and its oxides are not bioactive. Thus, no chemical bonding of the titanium implant with bone tissue occurs, which is beneficial for temporary implants, e.g., for fracture treatment. Figure 14d shows a Ti-plate for osteosynthesis which was treated by PEO to increase the thickness of the native titania layer. This serves to avoid contact welding with the titanium screws used for fixation and to allow unimpeded implant removal.

While, non-bioactivity of titania is beneficial for temporary implants, e.g., for fracture treatment, it is disadvantageous for permanent implants. Therefore, the surface of the implant has to be modified with bioactive coatings. The formation of a hydroxyapatite ($Ca_{10}(PO_4)_6(OH)_2$) coatings on titanium surfaces has been proven to facilitate osseointegration of implants without detrimental effects on the body. Generally, there are two ways to utilize PEO for the production of hydroxyapatite coatings: (1) Formation of a Ca- and P-containing titanium oxide coating by PEO and the subsequent hydrothermal treatment of the resulting coating to form hydroxyapatite (two-step process), or (2) formation of a hydroxyapatite-containing coating by PEO (one-step process). The most important factors for the bonding between implant and tissue are the composition of the coating, as well as its surface morphology and roughness [43,44].

Further medical applications for plasma electrolytic oxidation are shown in Figure 15 [42].

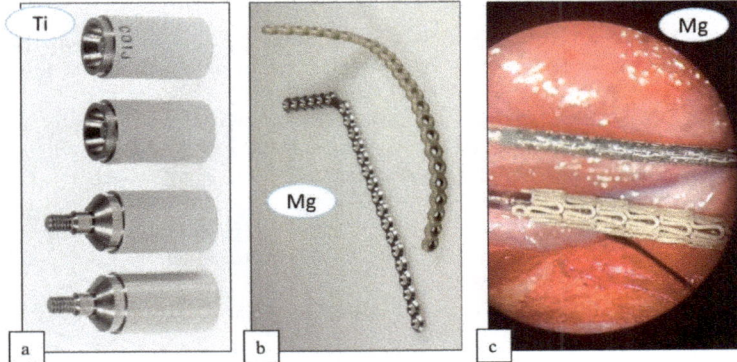

Figure 15. Application examples for PEO coatings on different substrates: scan body locators (**a**), plates for osteosynthesis before and after coating, (**b**) stent for treatment of coronary heart issues (**c**) products of Meotec GmbH [42].

The scan body locators, depicted in Figure 15a, are used in the dental implant technology to manufacture perfectly-fitting superstructures. Such prostheses are placed on posts anchored in the jaw and can support entire rows of teeth. The locators are deployed on laboratory analogs of impressions of the oral cavity and screwed in at the places where the implanted posts will later sit. The resulting geometry is then digitized using a three-dimensional (3D) laser scanner. The PEO coating on the locators (commercial name ScanOX®®) minimizes the optical reflection of the laser light and thus increases the quality of the data generated. Finally, the generated data are used for milling the superstructure, including perfectly positioned recesses for the jaw posts. Figure 15b,c shows a plate for osteosynthesis of bone fractures and a stent for treatment of coronal issues. The devices are made from magnesium, and they usually degrade within body tissue in a time-span of 12–24 months. The PEO coating (commercial name KERMASORB®®) regulates or delays this process so that the components remain stable in the first critical months [42].

6. Conclusions

Plasma electrolytic oxidation enables unique coating results due to its special process characteristics. Thanks to active research and development activities by scientific institutions and innovative companies, the selection of the treatable substrates as well as the composition and morphology of the layers that can be achieved have been expanded within wide limits. This allows for an increasing number of novel applications for the process in high technology fields. Recent advances in the field of process data analysis increasingly allow experimental access to the sub-mechanisms that occur during the PEO, and also form the basis for integrating the method into an increasingly digitized industry. Therefore, further progress in the field of plasma electrolytic oxidation can be expected in the next few years.

Funding: The financial support of this work by the Deutsche Forschungsgemeinschaft (DFG) is gratefully acknowledged (project number La-1274/55-1).

Acknowledgments: The support of Aalberts Surface Treatment GmbH, INNOVENT e.V. and Meotec GmbH by providing photographs of their products is grateful acknowledged. The support of Lisa-Marie Rymer and Morgan Uland is gratefully acknowledged as well.

Conflicts of Interest: The authors declare no conflict of interest.

References

1. Sluginov, N.P. On luminous phenomena, observed in liquids during electrolysis. *J. Russ. Phys. Chem. Soc.* **1880**, *12*, 193–203.
2. Güntherschulze, A.; Betz, H. *Electrolytic Capacitors*, 2nd ed.; Cram: Berlin, Germany, 1952.
3. Brown, S.D.; Kuna, K.J. Anodic Spark Deposition from Aqueous Soulutions of $NaAlO_2$ and Na_2SiO_3. *J. Am. Ceram. Soc.* **1971**, *54*, 385–390. [CrossRef]
4. Yerokhin, A.L.; Voevodin, A.A.; Lyubimov, V.V.; Zabinski, J.; Donley, M. Plasma electrolytic fabrication of oxide ceramic surface layers for tribotechnical purposes on aluminium alloys. *Surf. Coat. Technol.* **1998**, *110*, 140–146. [CrossRef]
5. Kurze, P. Production, Characterization and Application of Al_2O_3 Layers, Especially on Aluminium and Iron Materials, Dissertation. Ph.D. Thesis, TU Chemnitz, Chemnitz, Germany, 1981.
6. Krysmann, W.; Kurze, P.; Dittrich, K.-H.; Schneider, H.G. Process Characteristics and Parameters of Anodic Oxidation. *Cryst. Res. Technol.* **1984**, *19*, 973–979. [CrossRef]
7. Fattah-Alhosseini, A.; Molaei, M.; Kazem, B.; Babaei, K. Influence of Electrolyte Composition and Voltage on the Microstructure and Growth Mechanism of Plasma Electrolytic Oxidation (PEO) Coatings on Tantalum- A Review. *Anal. Bioanal. Electrochem.* **2020**, *12*, 517–535. [CrossRef]
8. Fattah-Alhosseini, A.; Babaei, K.; Molaei, M. Plasma electrolytic oxidation (PEO) treatment of zinc and its alloys. *Surf. Interfaces* **2020**, *18*, 100441. [CrossRef]
9. Krit, B.L.; Ludin, V.B.; Morozova, N.V.; Apelfeld, A.V. Microarc Oxidation of Carbon-Graphite Materials (Review). *Surf. Eng. Appl. Electrochem.* **2018**, *54*, 227–246. [CrossRef]
10. Lu, X.; Mohedano, M.; Blawert, C.; Matykina, E.; Arrabal, R.; Kainer, K.U.; Zheludkevich, M.L. Plasma electrolytic oxidation coatings with particle additions—A review. *Surf. Coat. Technol.* **2016**, *307*, 1165–1182. [CrossRef]
11. Fattah-Alhosseini, A.; Chaharmahali, R.; Babaei, K. Effect of particles addition to solution of plasma electrolytic oxidation (PEO) on the properties of PEO coatings formed on magnesium and its alloys. *J. Magnes. Alloys* **2020**. In Press. [CrossRef]
12. Sharifil, H.; Aliofkhazraei, M.; Darband, G.B.; Shrestha, S. A review on adhesion strength of PEO coatings by scratch test method. *Surf. Rev. Lett.* **2018**, *25*, 3. [CrossRef]
13. Gnedenkov, S.V.; Sinebryukhov, S.L.; Sergienko, V. Composite PEO-Coatings as Defence Against Corrosion and Wear: A Review. *Corros. Sci. Tech.* **2019**, *18*, 212–219. [CrossRef]
14. Clyne, T.W.; Troughton, S.C. A review of recent work on discharge characteristics during plasma electrolytic oxidation of various metals. *Int. Mater. Rev.* **2018**, *64*, 127–162. [CrossRef]
15. Tsai, D.S.; Chou, C.C. Review of the Soft Sparking Issues in Plasma Electrolytic Oxidation. *Metals* **2018**, *8*, 105. [CrossRef]
16. Walsh, F.C.; Low, C.T.J.; Wood, R.J.K.; Stevens, K.; Archer, J.; Poeton, A.R.; Ryder, A. Plasma electrolytic oxidation (PEO) for production of anodised coatings on lightweight metal (Al, Mg, Ti) alloys. *Trans. Inst. Met. Finish.* **2009**, *87*, 122–135. [CrossRef]
17. Yerokhin, A.L.; Nie, X.; Leyland, A.; Matthews, A.; Dowey, S.J. Plasma electrolysis for surface engineering. *Surf. Coat. Technol.* **1999**, *122*, 73–93. [CrossRef]
18. Sieber, M.; Simchen, F.; Morgenstern, R.; Scharf, I.; Lampke, T. Plasma Electrolytic Oxidation of High-Strength Aluminium Alloys—Substrate Effect on Wear and Corrosion Performance. *Metals* **2018**, *5*, 365. [CrossRef]
19. Sieber, M.; Mehner, T.; Dietrich, D.; Alisch, G.; Nickel, D.; Meyer, D.; Scharf, I.; Lampke, T. Wear-resistant coatings on aluminium produced by plasma anodising—A correlation of wear properties, microstructure, phase composition. *Surf. Coat. Technol.* **2014**, *240*, 96–102. [CrossRef]
20. Simchen, F.; Rymer, L.M.; Sieber, M. Composition of highly concentrated silicate electrolytes and ultrasound influencing the plasma electrolytic oxidation of magnesium. *IOP Conf. Ser. Mater. Sci. Eng.* **2017**, *181*, 012040. [CrossRef]
21. Sieber, M.; Simchen, F.; Scharf, I.; Lampke, T. Formation of a Spinel Coating on AZ31 Magnesium Alloy by Plasma Electrolytic Oxidation. *J. Mater. Eng. Perform.* **2016**, *25*, 1157–1162. [CrossRef]
22. Yerokhin, A.; Leyland, A.; Matthews, A. Kinetic aspects of aluminium titanate layer formation on titanium alloys by plasma electrolytic oxidation. *Appl. Surf. Sci.* **2002**, *200*, 172–184. [CrossRef]

23. Yerokhin, A.L.; Nie, X.; Leyland, A.; Matthews, A. Characterisation of oxide films produced by plasma electrolytic oxidation of a Ti–6Al–4V alloy. *Surf. Coat. Technol.* **2000**, *130*, 195–206. [CrossRef]
24. Cheng, Y.; Zhu, Z.; Zhang, Q.; Zhuang, X.; Cheng, Y. Plasma electrolytic oxidation of brass. *Surf. Coat. Technol.* **2020**, *385*, 125366. [CrossRef]
25. Malinovschi, V.; Marin, A.; Moga, S.; Negrea, D. Preparation and characterization of anticorrosive layers deposited by micro-arc oxidation on low carbon steel. *Surf. Coat. Technol.* **2014**, *253*, 194–198. [CrossRef]
26. Pezzato, L.; Brunelli, K.; Dolcet, P.; Dabalà, M. Plasma electrolytic oxidation coating produced on 39NiCrMo3 steel. *Surf. Coat. Technol.* **2016**, *307*, 73–80. [CrossRef]
27. Yang, W.; Li, Q.; Liu, W.; Liang, J.; Peng, Z.; Liu, B. Characterization and properties of plasma electrolytic oxidation coating on low carbon steel fabricated from aluminate electrolyte. *Vacuum* **2017**, *144*, 207–216. [CrossRef]
28. Yang, W.; Peng, Z.; Liu, B.; Liu, W.; Liang, J. Influence of Silicate Concentration in Electrolyte on the Growth and Performance of Plasma Electrolytic Oxidation Coatings Prepared on Low Carbon Steel. *J. Mater. Eng. Perform.* **2018**, *27*, 2345–2353. [CrossRef]
29. Ge, Y.; Wang, Y.; Cui, Y.; Zou, Y.; Guo, L.; Ouyang, J.; Jia, D.; Zhou, Y. Growth of plasma electrolytic oxidation coatings on Nb and corresponding corrosion resistance. *Appl. Surf. Sci.* **2019**, *491*, 526–534. [CrossRef]
30. Quintero, D.; Gómez, M.A.; Araujo, W.S.; Echeverría, F.; Calderón, J.A. Influence of the electrical parameters of the anodizing PEO process on wear and corrosion resistance of niobium. *Surf. Coat. Technol.* **2019**, *380*, 125067. [CrossRef]
31. He, S.; Ma, Y.; Ye, H.; Liu, X.; Dou, Z.; Xu, Q.; Wang, H.; Zhang, P. Ceramic oxide coating formed on beryllium by micro-arc oxidation. *Corros. Sci.* **2017**, *122*, 108–117. [CrossRef]
32. Cheng, Y.; Zhang, Q.; Zhu, Z.; Tu, W.; Cheng, Y.; Skeldon, P. Potential and morphological transitions during bipolar plasma electrolytic oxidation of tantalum in silicate electrolyte. *Ceram. Int.* **2020**, *46*, 13385–13396. [CrossRef]
33. Yao, Z.; Shen, Q.; Niu, A.; Hu, B.; Jiang, Z. Preparation of high emissivity and low absorbance thermal control coatings on Ti alloys by plasma electrolytic oxidation. *Surf. Coat. Technol.* **2014**, *242*, 146–151. [CrossRef]
34. Wang, L.; Zhou, J.; Liang, J.; Chen, J. Thermal control coatings on magnesium alloys prepared by plasma electrolytic oxidation. *Appl. Surf. Sci.* **2013**, *280*, 151–155. [CrossRef]
35. Available online: https://www.innovent-jena.de/ (accessed on 29 June 2020).
36. Kurze, P.; Krysmann, W.; Schreckenbach, J.; Schwarz, T.; Rabending, K. Coloured ANOF Layers on Aluminium. *Cryst. Res. Technol.* **1987**, *22*, 53–58. [CrossRef]
37. Wang, J.-M.; Tsai, D.-S.; Tsai, J.T.; Chou, C.-C. Coloring the aluminum alloy surface in plasma electrolytic oxidation with the green pigment colloid. *Surf. Coat. Technol.* **2017**, *321*, 164–170. [CrossRef]
38. Wang, S.; Liu, P. The technology of preparing green coating by conducting micro-arc oxidation on AZ91D magnesium alloy. *Pol. J. Chem. Technol.* **2016**, *18*, 36–40. [CrossRef]
39. Curran, J.A.; Kalkancı, H.; Magurova, Y.; Clyne, T.W. Mullite-rich plasma electrolytic oxide coatings for thermal barrier applications. *Surf. Coat. Technol.* **2007**, *201*, 8683–8687. [CrossRef]
40. Cengiz, S.; Uzunoglu, A.; Stanciu, L.; Tarakci, M.; Gencer, Y. Direct fabrication of crystalline hydroxyapatite coating on zirconium by single-step plasma electrolytic oxidation process. *Surf. Coat. Technol.* **2016**, *301*, 74–79. [CrossRef]
41. Antonio, R.F.; Rangel, C.E.; Mas, B.A.; Duek, E.A.R.; Cruz, N.C. Growth of hydroxyapatite coatings on tantalum by plasma electrolytic oxidation in a single step. *Surf. Coat. Technol.* **2019**, *357*, 698–705. [CrossRef]
42. Available online: http://www.meotec.eu/home/ (accessed on 29 June 2020).
43. Wang, Y.; Yu, H.; Chen, C.; Zhao, Z. Review of the biocompatibility of micro-arc oxidation coated titanium alloys. *Mater. Des.* **2015**, *85*, 640–652. [CrossRef]
44. Rafieerad, A.R.; Ashra, M.R.; Mahmoodian, R.; Bushroa, A.R. Surface characterization and corrosion behavior of calcium phosphate-base composite layer on titanium and its alloys via plasma electrolytic oxidation: A review paper. *Mater. Sci. Eng. C* **2015**, *85*, 640–652. [CrossRef]
45. Stojadinović, S.; Tadić, N.; Ćirić, A.; Vasilić, R. Photoluminescence properties of Eu^{3+} doped HfO_2 coatings formed by plasma electrolytic oxidation of hafnium. *Opt. Mater.* **2018**, *77*, 19–24. [CrossRef]
46. Stojadinovic, S.; Vasilić, R. Plasma electrolytic oxidation of hafnium. *Int. J. Refract. Met. Hard Mater.* **2017**, *69*, 153–157. [CrossRef]

47. Wang, J.; Li, C.; Yao, Z.; Yang, M.; Wang, Y.; Xia, Q.; Jiang, Z. Preparation of Fenton-like coating catalyst on Q235 carbon steel by plasma electrolytic oxidation in silicate electrolyte. *Surf. Coat. Technol.* **2016**, *307*, 1315–1321. [CrossRef]
48. Wang, J.; Yao, Z.; Yang, M.; Wang, Y.; Xia, Q.; Jiang, Z. A $Fe_3O_4/FeAl_2O_4$ composite coating via plasma electrolytic oxidation on Q235 carbon steel for Fenton-like degradation of phenol. *Environ. Sci. Pollut. Res. Int.* **2016**, *23*, 14927–14936. [CrossRef] [PubMed]
49. Lohrengel, M.M. Thin anodic oxide layers on aluminium and other valve metals: High field regime. *Mater. Sci. Eng. R.* **1993**, *11*, 243–294. [CrossRef]
50. Song, L.W.; Song, Y.W.; Shan, D.Y.; Zhu, D.Y.; Han, E.H. Product/metal ratio (PMR): A novel criterion for the evaluation of electrolytes on micro-arc oxidation (MAO) of Mg and its alloys. *Sci. China Technol. Sci.* **2011**, *54*, 2795–2801. [CrossRef]
51. Simchen, F.; Sieber, M.; Lampke, T. Electrolyte influence on ignition of plasma electrolytic oxidation processes on light metals. *Surf. Coat. Technol.* **2017**, *315*, 205–213. [CrossRef]
52. Ikonopisov, S. Theory of electrical breakdown during formation of barrier anodic films. *Electrochem. Acta* **1977**, *22*, 1077–1082. [CrossRef]
53. Yerokhin, A.L.; Snizhko, L.O.; Gurevina, N.L.; Leyland, A.; Pilkington, A.; Matthews, A. Discharge characterization in plasma electrolytic oxidation of aluminium. *J. Phys. D Appl. Phys.* **2003**, *36*, 2110–2120. [CrossRef]
54. Suminow, I.W.; Epelfeld, A.B.; Ljudin, W.B.; Krit, B.L.; Borisow, A.M. *Micro Arc Oxidation*; Ekomet: Moskow, Russia, 2005.
55. Matykina, E.; Arrabal, A.; Skeldon, P.; Thompson, G.E. Investigation of the growth processes of coatings formed by AC plasma electrolytic oxidation of aluminium. *Electrochem. Acta* **2009**, *54*, 6767–6778. [CrossRef]
56. Nie, X.; Meletis, E.I.; Jiang, J.C.; Leyland, A.; Yerokhin, A.L.; Matthews, A. Abrasive wear/corrosion properties and TEM analysis of Al_2O_3 coatings fabricated using plasma electrolysis. *Surf. Coat. Technol.* **2002**, *149*, 245–251. [CrossRef]
57. Morgenstern, R.; Sieber, M.; Lampke, T. Plasma electrolytic oxidation of AMCs. *IOP Conf. Ser. Mater. Sci. Eng.* **2016**, *118*. [CrossRef]
58. Oh, Y.O.; Mun, J.I.; Kim, J.W. Effects of alloying elements on microstructure and protective properties of Al_2O_3 coatings formed on aluminum alloy substrates by plasma electrolysis. *Surf. Coat. Technol.* **2009**, *240*, 141–148. [CrossRef]
59. Tillous, K.; Toll-Duchanoy, T.; Bauer-Grosse, E.; Hericher, L.; Geandier, G. Microstructure and phase composition of microarc oxidation surface layers formed on aluminium and its alloys. *Surf. Coat. Technol.* **2009**, *203*, 2969–2973. [CrossRef]
60. Schmeling, E.L.; Röschenbleck, B.; Weidemann, M.H. Process for the Production of Corrosion and Wear Resistant Protective Coatings on Magnesium and Magnesium Alloys. European Patent EP0333048A1, 15 March 1988.
61. Darband, G.B.; Aliofkhazraei, M.; Hamghalam, P.; Valizade, N. Plasma electrolytic oxidation of magnesium and its alloys: Mechanism, properties and applications. *J. Magnes. Alloys* **2017**, *5*, 74–132. [CrossRef]
62. Snizhko, L.O.; Yerokhin, A.L.; Gurevina, N.L.; Misnyankin, D.O.; Pilkington, A.; Leyland, A.; Matthews, A. A model for galvanostatic anodising of Al in alkaline solutions. *Electrochem. Acta* **2005**, *50*, 5458–5464. [CrossRef]
63. Snizhko, L.O.; Yerokhin, A.L.; Pilkington, A.; Gurevina, N.L.; Misnyankin, D.O.; Leyland, A.; Matthews, A. Anodic processes in plasma electrolytic oxidation of aluminium in alkaline solutions. *Electrochem. Acta* **2004**, *49*, 2085–2095. [CrossRef]
64. Dunleavy, C.S.; Golosnoy, I.O.; Curran, J.A.; Clyne, T.W. Characterisation of discharge events during plasma electrolytic oxidation. *Surf. Coat. Technol.* **2009**, *203*, 22. [CrossRef]
65. Alisch, G.; Nickel, D.; Lampke, T. Simultaneous plasma-electrolytic anodic oxidation (PAO) of Al-Mg compounds. *Surf. Coat. Technol.* **2011**, *206*, 1085–1090. [CrossRef]
66. Rakoch, A.G.; Gladkova, A.A.; Linn, Z.; Strekalina, D.M. The evidence of cathodic micro-discharges during plasma electrolytic oxidation of light metallic alloys and micro-discharge intensity depending on pH of the electrolyte. *Surf. Coat. Technol.* **2015**, *269*, 138–144. [CrossRef]
67. Guan, Y.; Xia, Y.; Li, G. Growth mechanism and corrosion behavior of ceramic coatings on aluminum produced by autocontrol AC pulse PEO. *Surf. Coat. Technol.* **2008**, *202*, 4602–4612. [CrossRef]

68. Timoshenko, A.V.; Magurova, Y.V. Application of oxide coatings to metals in electrolyte solutions by microplasma methods. *Rev. Metal.* **2000**, *36*, 323–330. [CrossRef]
69. Duan, H.; Li, Y.; Xia, Y.; Chen, S. Transient voltage-current characteristics: New insights. *Int. J. Electrochem. Sci.* **2012**, *7*, 7619–7630.
70. Rogov, A.B.; Shayapov, V.R. The role of cathodic current in PEO of aluminum. *Appl. Surf. Sci.* **2017**, *394*, 323–332. [CrossRef]
71. Rogov, A.B.; Matthews, A.; Yerokhin, A. Role of cathodic current in plasma electrolytic oxidation of Al. *Electrochim. Acta* **2019**, *317*, 221–231. [CrossRef]
72. Rogov, A.B.; Yerokhin, A.; Matthews, A. The Role of Cathodic Current in Plasma Electrolytic Oxidation of Aluminum. *Langmuir* **2017**, *33*, 11059–11069. [CrossRef]
73. Kazanski, B.; Kossenko, A.; Zinigrad, M.; Lugovskoy, A. Fluoride ions as modifiers of the oxide layer produced by plasma electrolytic oxidation on AZ91D magnesium alloy. *Appl. Surf. Sci.* **2013**, *287*, 461–466. [CrossRef]
74. Wang, L.; Chen, L.; Yan, Z.; Wang, H.; Peng, J. Effect of potassium fluoride on structure and corrosion resistance of plasma electrolytic oxidation films formed on AZ31 magnesium alloy. *J. Alloys Compd.* **2009**, *480*, 469–474. [CrossRef]
75. Nickel, D.; Alisch, G.; Händel, M.; Lampke, T.; Sieber, M. A Method of Treating a Voided Ceramic Protective Layer, with which a Substrate is Provided. DE 102014122451 B3, 29 August 2014.
76. Sealers for Duplex PEO Coatings. Available online: https://blog.keronite.com/sealers-for-plasma-electrolytic-oxidation (accessed on 28 June 2020).
77. Available online: https://www.aalberts-st.com/en/ (accessed on 28 June 2020).

© 2020 by the authors. Licensee MDPI, Basel, Switzerland. This article is an open access article distributed under the terms and conditions of the Creative Commons Attribution (CC BY) license (http://creativecommons.org/licenses/by/4.0/).

Article

Evaluation of the Corrosion Resistance and Cytocompatibility of a Bioactive Micro-Arc Oxidation Coating on AZ31 Mg Alloy

Shun-Yi Jian [1], Mei-Ling Ho [2,3,4,5], Bing-Ci Shih [6], Yue-Jun Wang [7], Li-Wen Weng [7], Min-Wen Wang [6,*] and Chun-Chieh Tseng [7,*]

1. Department of Chemistry and Materials Engineering, Chung Cheng Institute of Technology, National Defense University, Taoyuan 33591, Taiwan; ftvko@yahoo.com.tw
2. Department of Physiology, College of Medicine, Kaohsiung Medical University, Kaohsiung 80708, Taiwan; homelin@kmu.edu.tw
3. Orthopaedic Research Center, Kaohsiung Medical University, Kaohsiung 80708, Taiwan
4. Department of Marine Biotechnology and Resources, National Sun Yat-sen University, Kaohsiung 80424, Taiwan
5. Department of Medical Research, Kaohsiung Medical University Hospital, Kaohsiung 80708, Taiwan
6. Department of Mechanical Engineering and Graduate Institute of Mechanical and Precision Engineering, National Kaohsiung University of Applied Sciences, Kaohsiung 80778, Taiwan; sss10183@gmail.com
7. Combination Medical Device Technology Division, Medical Devices Department, Metal Industries Research & Development Centre, Kaohsiung 81160, Taiwan; fulick@mail.mirdc.org.tw (Y.-J.W.); only7110@mail.mirdc.org.tw (L.-W.W.)
* Correspondence: mwwang@nkust.edu.tw (M.-W.W); cctseng0915@gmail.com (C.-C.T.); Tel.: +886-7-6955298 (exit. 225) (C.-C.T.); Fax: +886-7-6955249 (C.-C.T.)

Received: 13 May 2019; Accepted: 14 June 2019; Published: 20 June 2019

Abstract: Magnesium alloys have recently been attracting attention as a degradable biomaterial. They have advantages including non-toxicity, biocompatibility, and biodegradability. To develop magnesium alloys into biodegradable medical materials, previous research has quantitatively analyzed magnesium alloy corrosion by focusing on the overall changes in the alloy. Therefore, the objective of this study is to develop a bioactive material by applying a ceramic oxide coating (magnesia) on AZ31 magnesium alloy through micro-arc oxidation (MAO) process. This MAO process is conducted under pulsed bipolar constant current conditions in a Si- and P-containing electrolyte and the optimal processing parameters in corrosion protection are obtained by the Taguchi method to design a coating with good anti-corrosion performance. The negative duty cycle and treatment time are two deciding factors of the coating's capability in corrosion protection. Microstructure characterizations are investigated by means of SEM and XRD. The simulation body-fluid solution is utilized for testing the corrosion resistance with the potentiodynamic polarization and the electrochemical impedance test data. Finally, an in vivo testing shows that the MAO-coated AZ31 has good cytocompatibility and anticorrosive properties.

Keywords: magnesium alloy; micro-arc oxidation; Taguchi method; SBF; in-vivo test; biodegradability

1. Introduction

Metallic materials are now widely used as biomaterials for applications in the field of orthopedics due to their high mechanical strength and fracture toughness. Compared to ceramics and polymeric materials, metallic materials are more suitable for load bearing applications [1]. The implant is usually in contact with the body fluids that typically have a high ionic intensity. It tends to induce the simultaneous electrochemical reaction between the surrounding fluids and the implanted metals. The corrosion of metals can release ions that may cause allergy, inflammation, diseases, or cancers [2–4].

Among the metallic biomaterials, the Young's modulus of 316 stainless steel (SS), cobalt-chromium (Co–Cr), and titanium (Ti) alloys are higher than that of human bone [5]. The difference in the Young's modulus between these materials and the bone tissues can induce stress-shielding effect [6]. Ti-based alloys have been considered once as the most promising metals for biomaterial applications due to their light weight, bio-inertness, and good corrosion resistance in the early 1970s [7]. Nevertheless, they are relatively poor in long-term tribological behaviors. Thus, the use of Ti-based alloys is limited in cases where good wear properties are required [8,9]. Most important of all, the Ti alloy screws used for bone fixation in the immature skeleton must be removed by a second surgical procedure after bone healing [4], which will cause patients pain and medical costs. In recent years, magnesium (Mg) and its alloys are emerging as the potential metallic materials for orthopedics due to their biodegradability in physiological body environment [5,10], excellent biocompatibility, and osteopromotion [5,11,12].

Unfortunately, Mg alloys still face some limitations as biodegradable medical materials. The most important issue concerning Mg alloy implants is that the Mg alloy corrodes too rapidly to maintain implant function before tissue healing using current technologies. Moreover, the corrosion process releases hydrogen and hydroxide ions, resulting in a substantial increase in the pH of the local area. These reactions might adversely affect local cell functions and delay the healing of surgical regions. The evolved hydrogen bubbles from a corroding magnesium implant can be accumulated in gas pockets next to the implant which can cause tissue necrosis and separation of tissues and tissue layers [13,14]. To overcome the challenges related to the use of Mg alloys, many studies have been working on finding ways to slow down the degradation of Mg alloys under physiological conditions in vitro through diverse approaches, such as development of new Mg alloys with different elements and deposition of anticorrosive coatings on Mg alloys substrates. Córdoba used a silane/TiO_2 coating to control the corrosion rate of Mg alloys in simulated body fluid [15]. Sachiko Hiromoto studied the self-healing properties of hydroxyapatite and octacalcium phosphate coatings on pure Mg and a Mg alloy [16]. Lei et al. enhanced the corrosion protection of MgO coatings on magnesium alloy deposited by an anodic electrodeposition process [17]. Among these coatings, micro-arc oxidation (MAO) coating has been demonstrated to be an effective coating for protection of Mg alloys [18,19].

In order to match the bone reconstruction and act as an effective biomechanical support, Witte et al. [20] reported that it is desirable to have the orthopedic implants present for at least 12 weeks to allow sufficient time for healing. However, the presence of a high pore density on the surface of the MAO coatings of Mg and its alloys reduces its corrosion protective ability [21]. Thus, many strategies have been utilized to improve the corrosion resistance of MAO-coated Mg alloys. Among them, it is recognized that optimization of the MAO control parameters, such as electrolytes, voltage, current density, and current mode, etc., to modify the microstructure and morphology is beneficial in the production of MAO-coated Mg alloy with high corrosion resistance [22–26]. It has been reported that the MAO coating formed using a bipolar current have a more compact structure with uniform coating thickness and fewer defects compared to the coating formed using a unipolar mode [26,27], resulting in the coating treated with the pulsed bipolar current mode had a great corrosion resistance than that of the MAO coating treated with the unipolar current mode. However, a systematic study about the electrical parameters for preparing corrosion resistant MAO coating using pulsed bipolar current mode has not been well documented in the literature. Therefore, in the present study, a dual electrolyte system consisting of Na_2SiO_3 and Na_3PO, as well as the addition of NaOH was used to produce MAO coatings under bipolar constant current mode. The influence of treatment time, negative duty cycle, current density, and pulsing frequency on the properties of MAO coatings was investigated. Our results showed that the novel Si- and P-containing MAO coating on AZ31 alloy prepared under the optimum electrical parameters found in this study has good cytocompatibility which was verified in in vivo tests.

2. Experimental

2.1. Preparation of Specimens

AZ31 magnesium alloy with size of 50 mm × 25 mm × 2 mm was used in this study as a substrate for MAO coating process. The AZ31 magnesium alloy has a chemical composition (wt %) of Al 2.5–3.5, Zn 0.5–1.5, Mn 0.2–0.5, Si ≤ 0.1, Cu ≤ 0.05, Fe ≤ 0.005%, and Mg balance. Prior to MAO coating, sample was mechanically polished using abrasive paper up to 2000 grit, degreased with acetone, rinsed with deionized water and dried in a stream of hot air of 60 °C. The treatment device for micro-arc oxidization process consists of a power generator, a 1 L glass container was used as an electrolytic cell, and as a stirring and cooling system. A pulse power generator (MIRDC) with work voltage up to 400 V, current up to 10 A, the duty cycle ranging from 5%–95%, and the electrical frequency ranging from 500–5000 Hz was used in the experiments for the formation of MAO coatings. The stainless steel plate and the Mg alloy substrate were used as the cathode and anode, respectively. The MAO treatments were conducted in the electrolyte containing 60 g/L Na_2SiO_3, 20 g/L Na_3PO_4, and 70 g/L NaOH. The solution stirred with a rate of 300 rpm was cooled to keep its temperature at 25 °C during the oxidation process. MAO treatments were performed using the pulsed bipolar current mode. The positive duty cycle was set at 40% and the negative current density magnitude equals to that of the positive one during the process.

The Taguchi method, a powerful tool for the experimental design of performance characteristics, is widely used to decide the optimal process parameters [28]. The strategy of experimental design used in the Taguchi method is based on orthogonal arrays and fractional factorial to estimate the effects of main factors on the process. In this respect, the Taguchi experimental design method can reduce the number of experiments while giving the full information of all the factors that affect the performance parameter. The first important step in design of experiment is the proper selection of factors and their levels. In this study, four operating factors include treatment time (10, 20, and 30 min), negative duty cycle (20%, 40%, and 60%), current density (50, 100, and 200 mA/cm^2), and pulsing frequency (500, 2000, and 4000 Hz) were considered in three levels. Thus, the experiment is based on the L9 orthogonal array in which the four control parameters are varied at three levels.

2.2. Characterization

The morphologies and compositions of the MAO coatings were detected by scanning electron microscopy (SEM, JSM-IT100, 15 kV, JEOL, Tokyo, Japan) with energy-dispersive X-ray analysis (EDX). The size and amount of micro-pores on the coating surface were counted quantitatively by Image-J software (open source version 1.43) from the SEM images. In addition, the crystallinity and phase of the coatings were characterized using X-ray diffraction (XRD; Philips powder X-ray diffractometer PW 1710, Amsterdam, Netherlands) with Cu Kα radiation (wavelength: 0.15405 nm). The roughness in the coating was analyzed by the ET400A (α-Step Talysurf, Sutronic 3+ profilometer, Taylor Hobson, Leicester, UK). The mean roughness (roughness average R_a) was used in this study. In hydrogen evolution method the amount of dissolved magnesium can be measured from the volume of hydrogen evolution as a result of the corrosion reaction [29–31]. Adhesion between the MAO coating and the substrate was determined by cross-cut test, following ASTM D-3359 [32] using 3M #600 tape.

2.3. Electrochemical Measurements

Potentiodynamic polarization test and electrochemical impedance spectroscopy (EIS) were employed to measure the corrosion resistance of the AZ31 substrate and MAO-coated samples in simulated body-fluid (SBF) Hanks solution using an Autolab PGSTAT30 potentiostat-frequency analyzer (Utrecht, Netherlands). This solution is usually employed in biological research projects attributable to the osmotic and ion concentrations that match those of the human body. Potentiodynamic polarization tests were performed in a three-electrode cell system in which a tested specimen of 1 cm^2, a platinum sheet and a saturated calomel electrode (SCE) were used as counter and reference electrodes, respectively. The potentiodynamic polarization curves were measured by sweeping the potential in the positive direction at a scan rate of 0.5 mV/s after a steady open circuit potential (OCP) was reached. The sweep range was from an initial potential of −300 mV to a final potential of 500 mV. Prior to the EIS measurement, each panel was immersed in the test solution for 30 min to reach a steady OCP. EIS data were obtained at the open circuit potential and ambient temperature with a voltage amplitude of 10 mV in the frequency range from 10^{-2}–10^5 Hz.

2.4. Animal Surgery and Implant Harvest

Animal experiments were conducted under the NIH guide for care and use of laboratory animals and approved by the animal ethics committee of Kaohsiung Medical University (no: IACUC-103052). The animal model used skeletally mature New Zealand White rabbits weighing 3.5–4.5 kg. All animals were kept in a single room and fed a dried diet and water ad libitum. Generally, an aesthesia was induced in all animals using intramuscular injections of ketamine 40 mg/kg and xylazine 10 mg/kg. One naked AZ31 and Si–P coated AZ31 screw samples were implanted into the femoral shaft of a rabbit. The rabbits were randomly divided into two groups of three each, designated AZ31 (untreated) and MAO (micro-arc oxidized AZ31). In each rabbit an identical pin was implanted into each of its femoral bones. The rabbits were euthanized humanely by using an intravenous overdose of barbiturate (200 mg/kg) after 4, 8, and 12 weeks post-implantation.

The degradation processes of AZ31 and MAO bone-screw samples were evaluated by micro-CT imagery (μCT, Skyscan 1272, Bruker, Kontich, Belgium). The scan conditions were performed at 12.0 μm scanning resolution, X-ray voltage of 100 kV, a current of 100 μA, exposure time of 2050 ms, and with a 0.11 mm × 2 mm copper filter. The rotation step was 0.5° per image and the averaging was 3 with a 360° scan. Reconstruction of sections was carried out with GPU-based scanner software (NRecon).

After CT scanning, the specimens were removed and prepared for histological examination. The specimens were fixed in 10% formalin solution, dehydrated, and embedded in methylmethacrylate. After polymerization, the specimens were stained and the images were then acquired using a camera coupled to a light microscope.

3. Results and Discussion

3.1. Determination of Optimal Level of Corrosion Resistance

3.1.1. Orthogonal Experiment and Analysis for MAO Process

In this study an attempt has been made to improve the corrosion resistance of the MAO-coated AZ31 Mg alloy by optimizing the MAO processing parameters using the Taguchi method. Figure 1 shows the signal-to-noise (S/N) response graph for the corrosion current density. Despite the objective function (corrosion current density) being a lower-the-better type of control function, the optimal level of the process parameters is the level with the highest S/N ratio. From these graphs it is clear that the optimum values of the factors and their levels are current density of 200 mA/cm^2, treatment time of 30 min, pulse frequency of 4000 Hz, and negative duty cycle of 20%.

Taguchi method introduced the following S/N ratio for evaluating the corrosion resistance performance of the system and the quality characteristic deviating from the desired values. It was clearly seen from Figure 1 that duty cycle and treatment time are two of the most important factors due to their higher contributions to reduce the corrosion. To confirm the optimum condition experimentally, additional experiments, such as varying the duty cycle and treatment time, were performed to verify the accuracy of the estimated results by the Taguchi method.

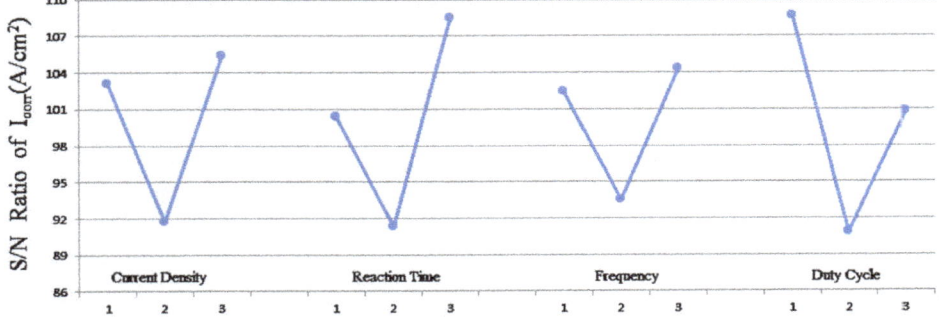

Figure 1. Effect of factors on corrosion current density (I_{corr}) of the coatings.

3.1.2. Confirmation Runs for the Negative Duty Cycle

In order to study the influence of the duty cycle on the corrosion behavior and physical properties of the MAO coatings, confirmation tests were conducted for 30 min under a constant current density of 200 mA/cm^2, a pulsing frequency of 4000 Hz, and three negative duty cycles of 10%, 20%, and 30%. The relevant results are listed in Table 1. The surface morphology of the MAO-coated AZ31 Mg alloy treated at negative duty cycle of 10%, 20% and 30%, respectively, is displayed in Figure 2a–c. A typical appearance of MAO coatings can be clearly seen from Figure 2. The surfaces of the coatings were dominated by many randomly distributed disc-like structures with open or sealed holes in the center as the result of the formation of discharge channels. SEM images reveal that the MAO-coated AZ31 alloy treated at the negative duty cycle of 20% has relatively less pores and smaller pore size compared to the MAO coatings formed at 10% and 30% negative duty cycles, which is consistent with [33]. The pore diameter showed a tendency to decrease with the increasing negative duty cycle as the negative duty cycle increases from 10% to 20%, while for the negative duty cycle is between 20% and 30%, an increase of pore diameter was evident. The decrease of pore diameter with the increasing negative duty cycle from 10% to 20% can be explained by the randomization of the anodic breakdown sites owing to the occurrence of cathodic breakdown during the cathodic period [34]. On the other hand, more cations will accumulate at the layer/electrolyte interface with increasing negative duty cycle from 20% to 30%, which might result in slightly stronger anodic breakdown and, thus, increasing the pore diameter. The SEM images shown in Figure 2 were image-processed by Image-J software for porosity analysis. The averaged porosities determined from on-surface SEM images are listed in Table 1. The MAO coating prepared with 20% negative duty exhibits the smallest porosity among these three coating as expected. The MAO-coated AZ31 alloy treated at the negative duty cycle of 20% was found to improve the coating quality in terms of surface morphology where the surface became smoother and the micro-pore size was reduced, which might influence the corrosion behavior. Nevertheless, the three duty cycles illustrate much alike surface morphology since such morphologies reflect the very last discharge events.

Cross-sectional SEM images of the coating were used to evaluate the thickness of the MAO layers. Figure 2d–f shows the cross-sections of the AZ31 plate after MAO treated with different duty cycles for 30 min. A very thick MAO layer with thickness about 24.2 μm was formed for the sample coated at 10% negative duty cycle (Figure 2d). The coating thicknesses decreased tremendously with the increase in negative duty cycle from 10% to 20%. A further increase of the negative duty cycle to 30%, the thickness of MAO coating was almost changeless, which are about 7.35 μm. Moreover, the sample prepared at the 30% negative duty cycle shows it has more porosity in the inner layer compared to sample prepared at the 20% negative duty cycle. It has been reported that the anode process promotes the coating growth, whereas the dissolution of some oxide phases on the coating surface takes place during the cathode process [35]. Obviously, the thickness of MAO coating will be governed by the relative rate of coating formation and dissolution. Since the anodic current density remained the same in the present study, it is likely that the increase of cathodic duty cycle promotes the dissolution of oxide layer, thus reducing the thickness of MAO coating. Moreover, the coating formed at the 20% negative duty cycle had a more compact microstructure in the inner layer compared with coatings prepared by the 10% and 30% negative duty cycles.

The results of the EDX analysis carried out on the surface of MAO coatings are given in Table 2. It can be seen from the table that EDX analysis of MAO coatings exhibits the presence of Mg, O, Al, Si, and P. The presence of Mg and Al elements implies that the substrate elements entered into the coating during the MAO process. Elements of Si and P came from the electrolyte. However, it shows the higher concentration of Mg and lower concentrations of Si and P for MAO formed at 20% negative duty cycle when compared with that of MAO formed at the 10% and 30% negative duty cycles. The surface morphology of MAO formed at the 20% negative duty cycle (Figure 2b) exhibited a lower pore density with a smaller pore diameter, which is expected to have a smaller surface area for the adsorption of silicate and phosphorous anions from the electrolyte, which resulted in lower silicon and phosphorous contents over the surface.

XRD spectra of all MAO-coated samples are given in Figure 3. There is no significant difference among these coatings. All the MAO coatings were mainly composed of MgO and $MgSiO_3$. MgO is formed via substrate metal oxidation during MAO treatment, while the $MgSiO_3$ phase is derived from the co-deposition of the alkaline electrolyte components into the coating structure. The $Mg_3(PO_4)_2$ possibly is existed in the coatings, however, no phosphorous-containing substances were identified by XRD. This might be ascribed to the amount of $Mg_3(PO_4)_2$ phase is low. The peaks of magnesium in XRD spectra for these three MAO coatings come from the AZ31 Mg alloy substrate, which indicate that the coating is thin.

To examine the adhesion of the MAO layers to the AZ31 Mg alloy substrates, adhesion tape tests according to ASTM D-3359 [32] were carried out. Our results showed that the adhesion of the samples could all be classified as 5B according to the ASTM D3359-17 [32] standard, indicating MAO is a promising surface treatment technology on magnesium alloys which provides good adhesion strength.

Potentiodynamic polarization test in SBF solution was carried out to evaluate the corrosion behavior of the MAO-coated AZ31 alloy (Figure 4). It can be seen that the sample treated at 20% negative duty cycle has more positive corrosion potential compared to other specimens. The corrosion current densities of the MAO coatings formed under negative duty cycle of 10%, 20%, and 30% are 2.23×10^{-6}, 8.05×10^{-7}, and 1.2×10^{-6} A/cm^2, respectively. Table 1 also presents the hydrogen evolution volume after an immersion time of 14 days. Among the three MAO samples treated at different negative duty cycles. It shows from Table 1 that the one treated at negative duty cycle of 20% exhibits the least hydrogen evolution volume (6 mL/14 day), whereas that treated at negative duty cycle of 30% has the most hydrogen evolution volume (16 mL/14 day), followed by the specimen treated at negative duty cycle of 10% (12 mL/14 day). The lower hydrogen evolution rate for MAO coating formed under the 20% negative duty cycle reveals that MAO sample formed under the 20% negative duty cycle can effectively prevent corrosive ions infiltrating into the interior of coating. Based on the above results, it is reasonable to suggest that the 20% negative duty cycle-treated MAO coating exhibits better corrosion resistance than the 10% negative duty cycle-treated specimen, even with a much thinner coating on the surface. This might be ascribed to less micro-pores inside the coating, decreasing the penetration of corrosive medium into the MAO coating and, thus, enhancing its corrosion resistance.

Table 1. Negative duty cycle effect on thickness, surface roughness, porosity, and corrosion current density of MAO-coated AZ31 Mg alloy.

Duty Cycle (%)	Thickness (μm)	Surface Roughness (μm)	Porosity (%)	Current Density (I_{corr} A/cm^2)	Hydrogen Volumes (mL/14 days)
10	24.2	0.45	4.65	2.23×10^{-6}	12
20	7.65	0.49	2.52	8.05×10^{-7}	6
30	7.35	1.10	3.57	1.2×10^{-6}	16

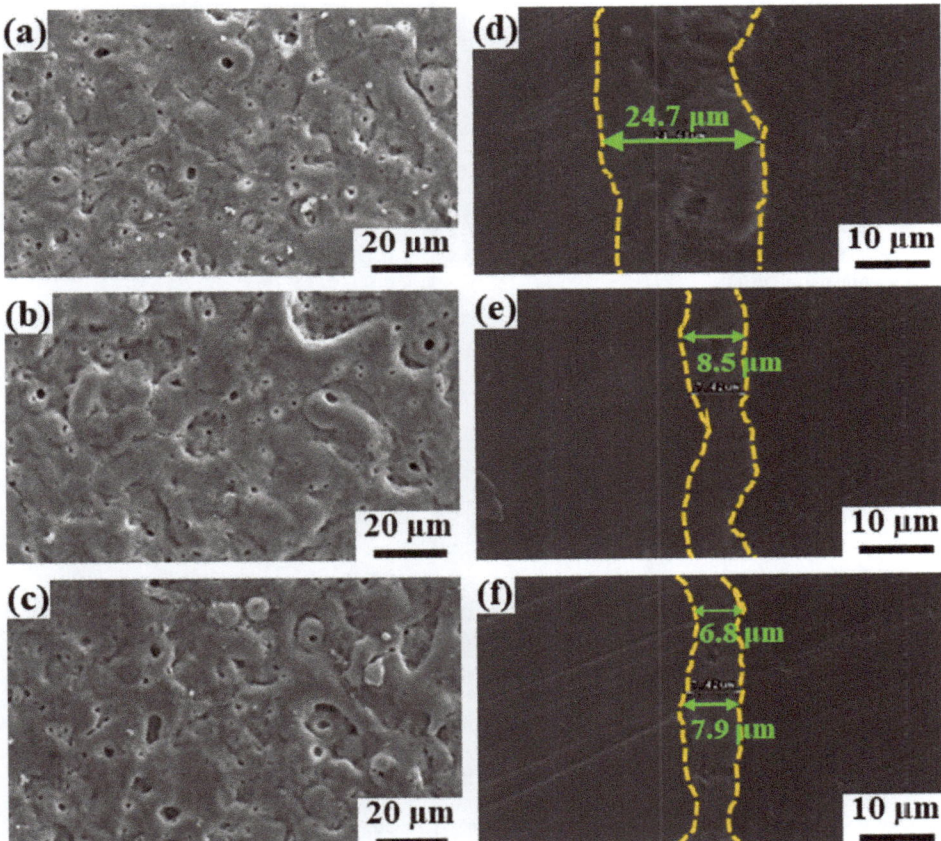

Figure 2. SEM-images of the surface (a–c) and cross-section (d–f) of the MAO coatings produced at negative duty cycle of 10% (a,d), 20% (b,e), and 30% (c,f), respectively, for 30 min.

Table 2. EDX analysis of the MAO coatings prepared at different negative duty cycles.

Duty Cycle (%)	Elements (wt %)					
	O	Mg	Al	Si	P	Ca
10	46.42	40.09	1.11	10.36	1.62	0.39
20	38.29	52.31	1.63	6.45	1.16	0.16
30	47.88	39.78	1.11	9.72	1.39	0.12

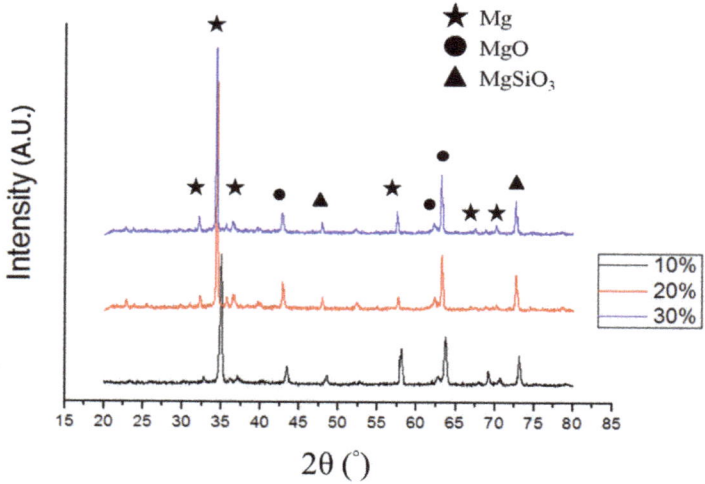

Figure 3. XRD patterns of oxide coatings formed at negative duty cycle of 10%, 20%, and 30%, respectively, for 30 min.

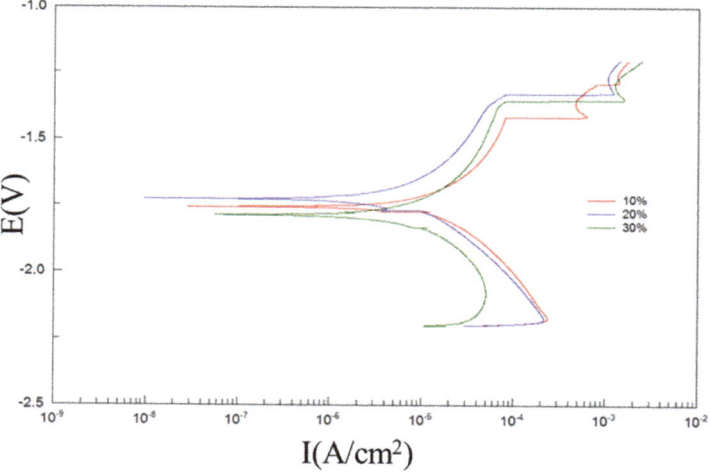

Figure 4. Potentiodynamic polarization curve of the MAO coating on the substrate surface formed at negative duty cycle of 10%, 20%, and 30%, respectively, for 30 min in the SBF.

3.1.3. Confirmation Runs for the Treatment Time

To compare the corrosion protection capabilities afforded by the MAO coatings prepared with different treatment times, a confirmation test was conducted at current density of 200 mA/cm^2, electrical frequency of 4000 Hz, and a negative duty cycle of 20% for 20, 30, and 40 min, respectively. The surface morphology of the MAO coatings produced with different oxidation times is displayed in Figure 5. Typical MAO porous microstructures were present on all three MAO coatings. The micro-porosity is considered to be "tracks" of the plasma discharge channels, through which the Mg plate and Mg ions from the substrate were likely discharged and reached the coating/electrolyte interface during the plasma-produced melting. The Mg and Mg ions then reacted with oxygen produced by electrolysis, and finally sintered and deposited on the coating surface, so providing to coating growth [36]. Therefore, the highest density of open (un-sealed) channels at the centers of the 'par cake' structures were observed on the surface of the 20-min MAO-coated sample (Figure 5a), which is a well-known feature of MAO coatings.

The cross-sectional microstructure of the coatings on AZ31 alloy after MAO processing for different oxidation times is shown in Figure 5d–f. Usually, when the duty ratio decreased, the thickness of the coating also increased due to the increasing treatment time. Figure 5d–f shows the thickness of the MAO coatings formed under treatment time. The coatings were 7.5 µm in 20 min, 8.5 µm in 30 min, and 10.8 µm in 40 min in thickness, respectively. It is demonstrated that the thickness is increasing when the treatment time increasing. Figure 5 also shows that the various samples contain of non-uniform surface oxide coating, the coating thickness is uneven caused by the violent reaction arc on the specimen's surface. Oxide film grows fast, but is not dense.

XRD patterns of the MAO coatings on AZ31 alloy after different treatment times are shown in Figure 6. All MAO coatings were mainly composed of MgO and $MgSiO_3$. It can be found that the peaks assigned to the substrate Mg decrease dramatically with increasing MAO treatment time, indicating that the coating gets thicker and more compact. In addition, the relative amount of the spinel $MgSiO_3$ phase to the MgO phase increases when the treatment time increases. This may be attributed to the more intense type B discharges prevailing at higher coating thickness. An increase in the intensity and temperature of the micro-arc discharges results in the increase of the reactive activity of the electrolyte component.

The potentiodynamic polarization curves of MAO samples produced with different treatment times in SBF solution are shown in Figure 7. From the curves shown in Figure 7, it can be observed that the sample treated for 30 min has more positive corrosion potential compared to other specimens. The corrosion current density of the sample treated for 30 min was 8.05×10^{-7} A/cm^2, lower than other specimens. This indicates the corrosion behavior of the MAO coating is highly dependent on the treatment time used for the coating formation. The cathodic polarization curve might be thought of as an indication of hydrogen evolution caused by water reduction. The kinetics of the cathodic reaction in the sample treated for 30 min are slower compared with other specimens. This phenomenon indicates that the cathodic reaction was kinetically more difficult in the sample treated for 30 min. The value of I_{corr} and E_{corr} are listed in Table 3, together with thickness, roughness, porosity, and hydrogen evolution volume. From the Table 3, it can be deduced that the noblest corrosion potential and the lowest corrosion current density were attained from the sample treated for 30 min, indicating that this sample corroded slower than other specimens. In conclusion, the corrosion resistance of MAO-coated AZ31 alloy is enhanced with increasing treatment time up to 30 min, while a further increase resulted in the reverse.

Table 3. Treatment time effect on thickness, surface roughness, porosity, and corrosion current density of MAO-coated AZ31 Mg alloy.

Treatment Time (min)	Thickness (µm)	Surface Roughness (µm)	Porosity (%)	Corrosion Current Density (I_{corr} A/cm^2)	Hydrogen Volumes (mL/14 days)
20	7.50	1.21	3.88	3.06×10^{-6}	10
30	7.65	0.49	2.52	8.05×10^{-7}	6
40	10.8	0.68	3.65	1.13×10^{-6}	7

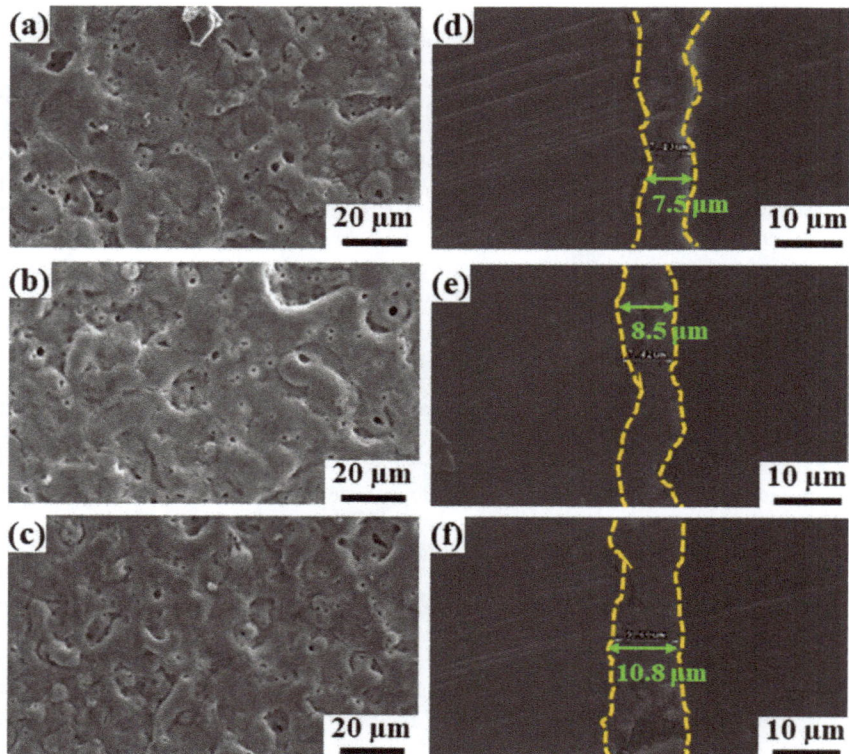

Figure 5. SEM-images of the surface (**a**–**c**) and cross-section (**d**–**f**) of the MAO coatings produced at negative duty cycle of 20% for treatment times of 20 min (**a**,**d**), 30 min (**b**,**e**), and 40 min (**c**,**f**).

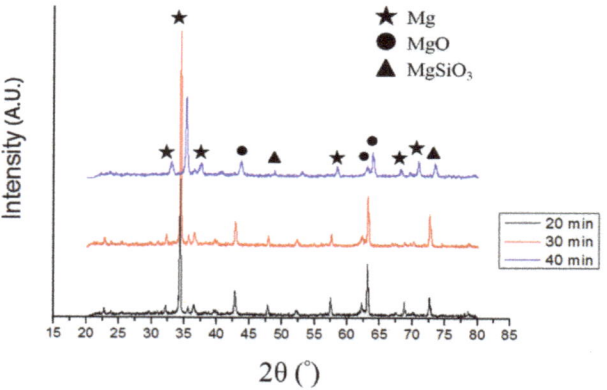

Figure 6. XRD patterns of oxide coatings formed at negative duty cycle of 20% for treatment times of 20, 30, and 40 min.

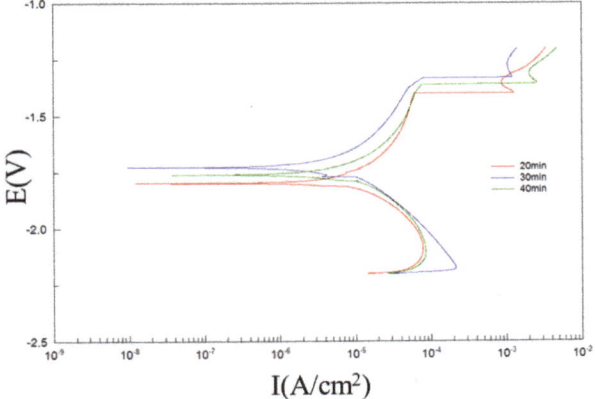

Figure 7. Potentiodynamic polarization curve of the MAO coating on the substrate surface formed at negative duty cycle of 20% for treatment times of 20, 30, and 40 min in the SBF.

3.2. Corrosion Resistance of Electrochemical Analysis

By means of the analyzed level effect, and also analysis of variance in the Taguchi method, we can derive the optimal conditional combination: current density at 200 mA/cm^2, electrical frequency at 4000 Hz, duty cycle at 20%, and treatment time of 30 min. After all, the results of the optimal setting showed that the S/N ratios obtained by the optimal results all fall in the 95% dependence interval. This study illustrates that the verification runs prove the reliability in the Taguchi method.

Consequently, in order to further explain the long-term corrosion protection performance of the MAO coatings, it is necessary to obtain a thorough clarification of their corrosion process and the combined electrochemical behaviors. The corrosion properties of the AZ31 substrate and the optimal conditional-coated AZ31 plates were evaluated using a potentiodynamic polarization test and it was conducted after immersion in SBF solution for different time periods of up to seven days. The corrosion current density (I_{corr}) of the different samples as a function of time is plotted in Figure 8. From this figure, it is apparent that the MAO-coated AZ31 plates had a lesser value of I_{corr} than the AZ31 substrate in all cases. It can be proposed that the coating formed with the optimal process conditions provides better corrosion protection for the AZ31 Mg alloy.

The corrosion resistance of bare AZ31 substrate and MAO-coated AZ31 obtained under the optimal conditions was further studied using EIS (Figure 9). The Nyquist plot of both samples exhibited a capacitive loop at high and medium frequencies and an inductive loop at low frequencies. It ought to be noted that, in spite of the Nyquist plot form, the sample with the MAO coating exhibits much larger semicircles compared with the uncoated AZ31 alloy sample, suggesting a far better anti-corrosive property compared to that of bare AZ31. It had been expected that the observed micro-pores formed on the MAO coating would have a negative effect on the corrosion performance. However, such an effect wasn't detected due to the compact inner part of the MAO treatment that provides effective anti-corrosion behavior beneath the outer porous region. Accordingly, the coated substrate exhibited a better corrosion resistance than the uncoated one based on the electrochemical tests.

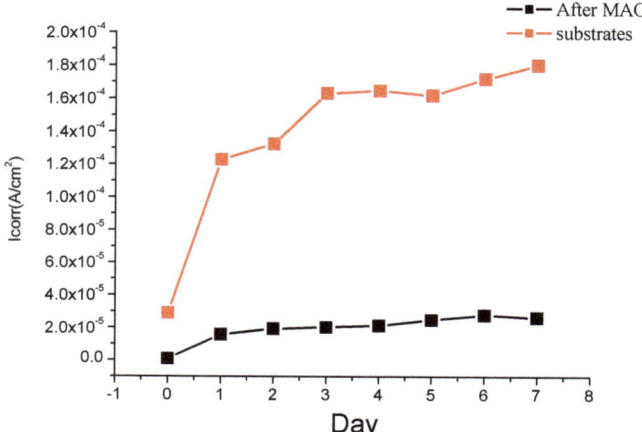

Figure 8. Corrosion current density of the bare and MAO-coated AZ31 Mg alloy immersed in SBF solution for different times.

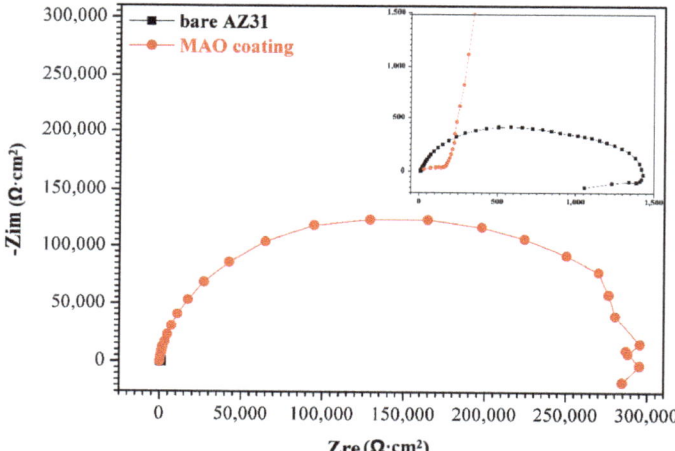

Figure 9. Impedance spectra of bare and MAO-coated AZ31 Mg alloy in the SBF after immersion of 7 days.

3.3. Bio-Degradable Implantation Test

In vitro and in vivo studies are necessary to evaluate the biocompatibility and osteoconductivity of these biomaterials before clinical applications. It is worth mentioning that the sample has already passed ISO10993 [37] including cytotoxicity tests, irritation tests, and sensitization tests, which were examined by SGS Taiwan Ltd. (the related proofs are provided in the supplementary materials). To better mimic the performance of MAO-coated Mg alloys in physiological environments, the in vivo implantation test was thus employed in this study. Figure 10 shows the result of the implant surgery after the magnesium bone-screw was implanted in the femora of a mature rabbit. The untreated AZ31 and MAO screws were implanted in vivo as a load-bearing implant to fix a bone fracture and, subsequently, qualitative micro-computed tomography (μCT) analysis and histological analysis were performed respectively at 4, 8, and 12 weeks. μCT scanning was performed to analyze the in vivo degradation of the Mg-based screw, new bone formation around the implant, and the fracture healing process. The representative μCT images of femora with screws at 4, 8 and 12 weeks postoperation were shown in Figure 11. The μCT sequels revealed the preliminary results of the bio-degradation of the untreated AZ31screw (Figure 11A–C) and MAO screw (Figure 11D–F) in the rabbit tissue cells. After 4 weeks implantation, it can be clearly seen from Figure 11A that shadows appeared on the surface of the AZ31 bone-screw, suggesting the corrosion reaction took place between the cell and bone-screw. There was a great decrease in AZ31 screw volume after 12 weeks of implantation (Figure 11C). In addition, bone tissue surrounded screw was observed after 12 weeks. The MAO-coated AZ31, in contrast, had positive influence in bio-degradation test. The bone-screw maintained a bright contrast in the μCT sequels as showed in Figure 11D–F). The MAO screw maintained the original morphology within 12 weeks, whereas AZ31 screw suffered obvious corrosion that several pits present on the surface was observed within 4 weeks. The analysis also revealed a slight decrease in the shrinking dimension of MAO screw compared to AZ31 screw after 12 weeks implantation. According to μCT results, AZ31 and MAO-coated AZ31 exhibited different degradation behaviors. AZ31 alloy degraded quickly, whereas MAO-coated AZ31 showed a slow degradation rate.

It is known that the degradation process was accompanied by hydrogen gas formation [38]. The large amounts of gas accelerate the degradation of the implant, on one hand, and interfere with the bone growth on the other [39]. With the lower degradation rate and correspondingly reduced gas formation, good healing results should be achieved when bones are implanted with MAO screws. This will be shown further by examining the interface between extra bone-screws and bone cells. A series of pathological photographs in Figure 12 provides examples of the extra bone-screw/bone cell interface of AZ31 and MAO screws at each time point. The inner part displayed in Figure 12 showed the extra bone screw and the outer part was the bone cell tissue. As can be seen in Figure 12A–C, the inner part reduced continuously from week 4 to week 12, indicating corrosion progress in the AZ31 screw by week 12. In contrast, the MAO screw was still visible after 12 weeks and the newly formed bones are direct contact to the screw (Figure 12D–F). This can be ascribed to its moderate degradation rate. These observations suggest that the degradation rate of the Mg alloy-based screw can impact its performance, which is consistent with the literature [40]. The AZ31 screw might have a degradation rate which is faster than the healing rate of a bone fracture, and the screw will degrade away before the healing process is over. Our results indicated that the novel Si-P MAO coating had positive effects not only on the bio-degradable results, but also on the bio-compatible trends.

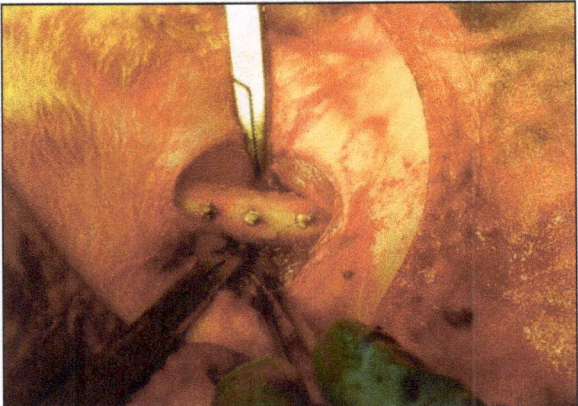

Figure 10. The surgery image of the implantation of the magnesium bone-screw in the femora of a mature rabbit.

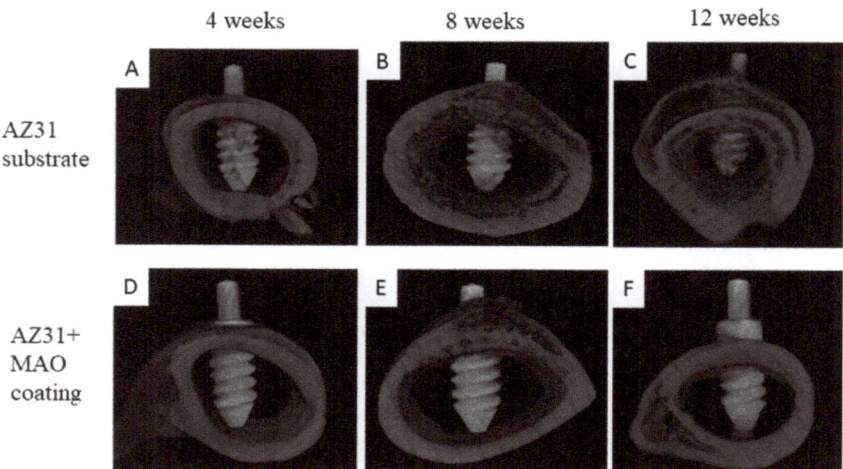

Figure 11. The μCT reconstructed images showing the degradation processes of AZ31 bone-screw (**A–C**) and MAO coating (**D–F**) after implantation of 4, 8, and 12 weeks.

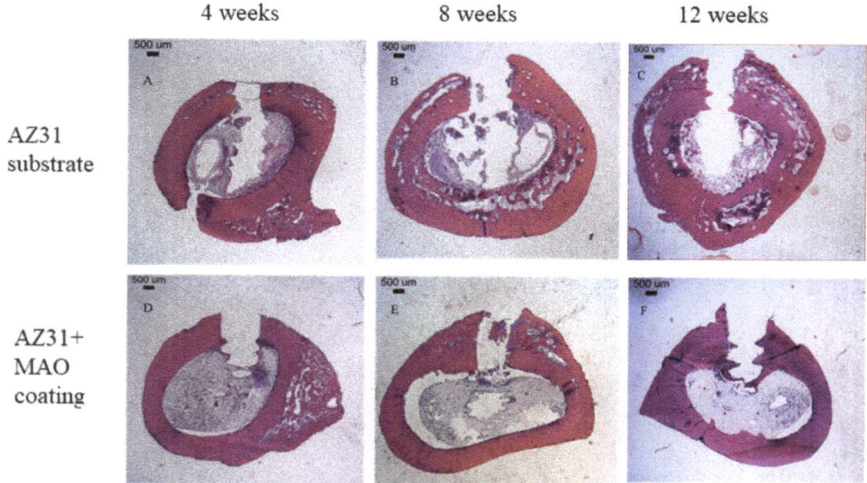

Figure 12. Pathological photographs of the screw/bone interfaces after AZ31 substrate (**A–C**) and the MAO coating (**D–F**) implanted in a mature rabbit for 4, 8, and 12 weeks.

4. Conclusions

In summary, the main findings of the work are:

(1). This study used the Taguchi method to find out the optimal conditional combination to produce the micro arc oxidation coating on AZ31 Mg alloy with the greatest corrosion resistance. The corrosion resistance of different MAO-coated AZ31 substrates was evaluated by electrochemical measurements carried out in SBF solution. Accordingly, the MAO-coated substrate exhibited a better corrosion resistance than the uncoated one based on the electrochemical tests.

(2). The optimal conditional combination obtained in this study is: current density of 200 mA/cm^2, electrical frequency of 4000 Hz, duty cycle of 20%, and treatment time of 30 min.

(3). The surface chemical components were analyzed by XRD. The results demonstrated that all the MAO coatings were mainly composed of MgO and $MgSiO_3$. Then, an in vivo testing showed that the MAO-coated AZ31 had good cytocompatibility and anticorrosive properties.

(4). In vivo tests showed that the degradation rate of the Mg alloy-based screw is critical to the success of its application. The well-adhered MAO layer on AZ31 Mg alloy efficiently reduces the degradation rate of the screw. This is beneficial to bone healing.

Supplementary Materials: The following are available online at http://www.mdpi.com/2079-6412/9/6/396/s1.

Author Contributions: Conceptualization, C.-C.T.; Data Curation, B.-C.S., Y.-J.W. and L.-W.W.; Project Administration, M.-W.W. and M.-L.H.; Writing—Original Draft, S.-Y.J.

Funding: This research was funded by Department of Industrial Technology, Ministry of Economic Affair, Taiwan.

Conflicts of Interest: The authors declare no conflict of interest.

References

1. Ratner, B.D.; Hoffman, A.S.; Schoen, F.J.; Lemons, J.E. *Biomaterials Science: An Introduction to Materials in Medicine*, 3rd ed.; Elsevier: Amsterdam, The Netherlands, 2012.
2. Jacobsen, N.; Hensten-Pettersen, A. Occupational health problems and adverse patient reactions in orthodontics. *Eur. J. Orthod.* **1989**, *11*, 254–264. [CrossRef] [PubMed]
3. Sun, Z.L.; Wataha, J.C.; Hanks, C.T. Effects of metal ions on osteoblast-like cell metabolism and differentiation. *J. Biomed. Mater. Res.* **1997**, *34*, 29–37. [CrossRef]
4. Vahey, J.W.; Simonian, P.T. Carcinogenicity and metallic implants. *Am. J. Orthop.* **1995**, *24*, 319–324. [PubMed]
5. Zhao, D.; Witte, F.; Lu, F.; Wang, J.; Li, J.; Qin, L. Current status on clinical applications of magnesium-based orthopaedic implants: A review from clinical translational perspective. *Biomaterials* **2017**, *112*, 287–302. [CrossRef] [PubMed]
6. Huiskes, R.; Weinans, H.; Van Rietbergen, B. The relationship between stress shielding and bone resorption around total hip stems and the effects of flexible materials. *Clin. Orthop. Relat. Res.* **1992**, *274*, 124–134. [CrossRef]
7. Long, M.; Rack, H.J. Titanium alloys in total joint replacement—A materials science perspective. *Biomaterials* **1998**, *19*, 1621–1639. [CrossRef]
8. Agins, H.J.; Alcock, N.W.; Bansal, M.; Salvati, E.A.; Wilson, P.D.; Pellicci, P.M.; Bullough, P.G. Metallic wear in failed titanium-alloy total hip replacements. A histological and quantitative analysis. *J. Bone Joint Surg.* **1988**, *70*, 347–356. [CrossRef]
9. Dearnley, P.A.; Dahm, K.L.; Çimenoğlu, H. The corrosion—Wear behaviour of thermally oxidised CP-Ti and Ti–6Al–4V. *Wear* **2004**, *256*, 469–479. [CrossRef]
10. Wu, G.; Ibrahim, J.M.; Chu, P.K. Surface design of biodegradable magnesium alloys—A review. *Surf. Coat. Technol.* **2013**, *233*, 2–12. [CrossRef]
11. Minkowitz, R.B.; Bhadsavle, S.; Walsh, M.; Egol, K.A. Removal of painful orthopaedic implants after fracture union. *J. Bone Joint Surg. Am.* **2007**, *89*, 1906–1912.
12. Zreiqat, H.; Howlett, C.R.; Zannettino, A.; Evans, P.; Schulze-Tanzil, G.; Knabe, C.; Shakibaei, M. Mechanisms of magnesium-stimulated adhesion of osteoblastic cells to commonly used orthopaedic implants. *J. Biomed. Mater. Res.* **2002**, *62*, 175–184. [CrossRef] [PubMed]
13. Kirkland, N.T.; Birbilis, N. *Magnesium Biomaterials: Design, Testing, and Best Practice*; Springer: Cham, Switzerland, 2014.
14. Song, G.; Song, S.Z. A Possible biodegradable magnesium implant material. *Adv. Eng. Mater.* **2007**, *9*, 298–302. [CrossRef]
15. Córdoba, L.C.; Montemor, M.F.; Coradin, T. Silane/TiO_2 coating to control the corrosion rate of magnesium alloys in simulated body fluid. *Corros. Sci.* **2016**, *104*, 152–161. [CrossRef]
16. Hiromoto, S. Self-healing property of hydroxyapatite and octacalcium phosphate coatings on pure magnesium and magnesium alloy. *Corros. Sci.* **2015**, *100*, 284–294. [CrossRef]
17. Lei, T.; Ouyang, C.; Tang, W.; Li, L.F.; Zhou, L.S. Enhanced corrosion protection of MgO coatings on magnesium alloy deposited by an anodic electrodeposition process. *Corros. Sci.* **2010**, *52*, 3504–3508. [CrossRef]

18. Cai, Q.; Wang, L.; Wei, B.; Liu, Q. Electrochemical performance of microarc oxidation films formed on AZ91D magnesium alloy in silicate and phosphate electrolytes. *Surf. Coat. Technol.* **2006**, *200*, 3727–3733. [CrossRef]
19. Arrabal, R.; Matykina, E.; Viejo, F.; Skeldon, P.; Thompson, G.E. Corrosion resistance of WE43 and AZ91D magnesium alloys with phosphate PEO coatings. *Corros. Sci.* **2008**, *50*, 1744–1752. [CrossRef]
20. Witte, F.; Kaese, V.; Haferkamp, H.; Switzer, E.; Meyer-Lindenberg, A.; Wirth, C.J.; Windhagen, H. In vivo corrosion of four magnesium alloys and the associated bone response. *Biomaterials* **2005**, *26*, 3557–3563. [CrossRef]
21. Narayanan, T.S.; Park, I.S.; Lee, M.H. Strategies to improve the corrosion resistance of microarc oxidation (MAO) coated magnesium alloys for degradable implants: Prospects and challenges. *Prog. Mater. Sci.* **2014**, *60*, 1–71. [CrossRef]
22. Mori, Y.; Koshi, A.; Liao, J.; Asoh, H.; Ono, S. Characteristics and corrosion resistance of plasma electrolytic oxidation coatings on AZ31B Mg alloy formed in phosphate—Silicate mixture electrolytes. *Corros. Sci.* **2014**, *88*, 254–262. [CrossRef]
23. Yue, Y.A.N.G.; Hua, W.U. Effect of current density on corrosion resistance of micro-arc oxide coatings on magnesium alloy. *T. Nonferr. Metal. Soc. China* **2010**, *20*, s688–s692.
24. Liu, J.; Zhang, W.; Zhang, H.; Hu, X.; Zhang, J. Effect of microarc oxidation time on electrochemical behaviors of coated bio-compatible magnesium alloy. Strategies to improve the corrosion resistance of microarc oxidation (MAO) coated magnesium alloys for degradable implants: Prospects and challenges. *Mater. Today Proc.* **2014**, *1*, 70–81. [CrossRef]
25. Gu, Y.; Chen, C.F.; Bandopadhyay, S.; Ning, C.; Zhang, Y.; Guo, Y. Corrosion mechanism and model of pulsed DC microarc oxidation treated AZ31 alloy in simulated body fluid. *Appl. Surf. Sci.* **2012**, *258*, 6116–6126. [CrossRef]
26. Hussein, R.O.; Zhang, P.; Nie, X.; Xia, Y.; Northwood, D.O. The effect of current mode and discharge type on the corrosion resistance of plasma electrolytic oxidation PEO) coated magnesium alloy AJ62. *Surf. Coat. Technol.* **2011**, *206*, 1990–1997. [CrossRef]
27. Hussein, R.O.; Nie, X.; Northwood, D.O. Influence of process parameters on electrolytic plasma discharging behaviour and aluminum oxide coating microstructure. *Surf. Coat. Technol.* **2010**, *205*, 1659–1667. [CrossRef]
28. Ben-Arfa, B.A.; Salvado, I.M.M.; Frade, J.R.; Pullar, R.C. Fast route for synthesis of stoichiometric hydroxyapatite by employing the Taguchi method. *Mater. Des.* **2016**, *109*, 547–555. [CrossRef]
29. Song, G. Control of biodegradation of biocompatable magnesium alloys. *Corros. Sci.* **2007**, *49*, 1696–1701. [CrossRef]
30. Atrens, A.; Liu, M.; Abidin, N.I.Z. Corrosion mechanism applicable to biodegradable magnesium implants. *Mater. Sci. Eng. B* **2011**, *176*, 1609–1636. [CrossRef]
31. Neil, W.C.; Forsyth, M.; Howlett, P.C.; Hutchinson, C.R.; Hinton, B.R.W. Corrosion of heat treated magnesium alloy ZE41. *Corros. Sci.* **2011**, *53*, 3299–3308. [CrossRef]
32. ASTM D-3359. *Standard Test Methods for Rating Adhesion by Tape Test*; ASTM International: West Conshohocken, PA, USA, 2017.
33. Lee, S.J.; Do Toan, L.H.; Lee, J.L.; Chen, C.Y.; Peng, H.C. Effects of pulsed unipolar and bipolar current regimes on the characteristics of micro-arc oxidation coating on LZ91 magnesium-lithium alloy. *Int. J. Electrochem. Sci.* **2018**, *13*, 2705–2717. [CrossRef]
34. Sah, S.P.; Tsuji, E.; Aoki, Y.; Habazaki, H. Cathodic pulse breakdown of anodic films on aluminium in alkaline silicate electrolyte—Understanding the role of cathodic half-cycle in AC plasma electrolytic oxidation. *Corros. Sci.* **2012**, *55*, 90–96. [CrossRef]
35. Yerokhin, A.L.; Nie, X.; Leyland, A.; Matthews, A. Characterisation of oxide films produced by plasma electrolytic oxidation of a Ti–6Al–4V alloy. *Surf. Coat. Technol.* **2000**, *130*, 195–206. [CrossRef]
36. Hussein, R.O.; Nie, X.; Northwood, D.O. An investigation of ceramic coating growth mechanisms in plasma electrolytic oxidation (PEO) processing. *Electrochim. Acta* **2013**, *112*, 111–119. [CrossRef]
37. ISO 10993-1. *Standard Test Methods for Biological Evaluation of Medical Devices-Part 1: Evaluation and Testing within a Risk Management Process*; ISO: Geneva, Switzerland, 2018.
38. Wilke, B.M.; Zhang, L.; Li, W.; Ning, C.; Chen, C.F.; Gu, Y. Corrosion performance of MAO coatings on AZ31 Mg alloy in simulated body fluid vs. Earle's balance salt solution. *Appl. Surf. Sci.* **2016**, *363*, 328–337. [CrossRef]

39. Waizy, H.; Diekmann, J.; Weizbauer, A.; Reifenrath, J.; Bartsch, I.; Neubert, V.; Schavan, R.; Windhagen, H. In vivo study of a biodegradable orthopedic screw (MgYREZralloy) in a rabbit model for up to 12 months. *J. Biomater. Appl.* **2014**, *28*, 667–675. [CrossRef] [PubMed]
40. Zhang, N.; Zhao, D.; Liu, N.; Wu, Y.; Yang, J.; Wang, Y.; Xie, H.; Ji, Y.; Zhou, C.; Zhuang, J.; et al. Assessment of the degradation rates and effectiveness of different coated Mg-Zn-Ca alloy scaffolds for in vivo repair of critical-size bone defects. *J. Mater. Sci. Mater. Med.* **2018**, *29*, 138. [CrossRef]

© 2019 by the authors. Licensee MDPI, Basel, Switzerland. This article is an open access article distributed under the terms and conditions of the Creative Commons Attribution (CC BY) license (http://creativecommons.org/licenses/by/4.0/).

Article

Degradation Behaviour of Mg0.6Ca and Mg0.6Ca2Ag Alloys with Bioactive Plasma Electrolytic Oxidation Coatings

Lara Moreno [1,*], Marta Mohedano [1], Beatriz Mingo [2], Raul Arrabal [1] and Endzhe Matykina [1]

1. Departamento de Ingeniería Química y de Materiales, Facultad de Ciencias Químicas, Universidad Complutense, 28040 Madrid, Spain; mmohedan@ucm.es (M.M.); raularrabal@quim.ucm.es (R.A.); ematykin@ucm.es (E.M.)
2. School of Materials, The University of Manchester, Oxford Road, Manchester M13 9PL, UK; beatriz.mingo@manchester.ac.uk
* Correspondence: laramo01@ucm.es

Received: 14 May 2019; Accepted: 11 June 2019; Published: 13 June 2019

Abstract: Bioactive Plasma Electrolytic Oxidation (PEO) coatings enriched in Ca, P and F were developed on Mg0.6Ca and Mg0.6Ca2Ag alloys with the aim to impede their fast degradation rate. Different characterization techniques (SEM, TEM, EDX, SKPFM, XRD) were used to analyze the surface characteristics and chemical composition of the bulk and/or coated materials. The corrosion behaviour was evaluated using hydrogen evolution measurements in Simulated Body Fluid (SBF) at 37 °C for up to 60 days of immersion. PEO-coated Mg0.6Ca showed a 2–3-fold improved corrosion resistance compared with the bulk alloy, which was more relevant to the initial 4 weeks of the degradation process. In the case of the Mg0.6Ag2Ag alloy, the obtained corrosion rates were very high for both non-coated and PEO-coated specimens, which would compromise their application as resorbable implants. The amount of F^- ions released from PEO-coated Mg0.6Ca during 24 h of immersion in 0.9% NaCl was also measured due to the importance of F^- in antibacterial processes, yielding 33.7 µg/cm^2, which is well within the daily recommended limit of F^- consumption.

Keywords: corrosion; PEO; biodegradable implants

1. Introduction

Over the last decades, magnesium and its alloys have been extensively investigated due to their excellent properties that make them promising candidates for biodegradable orthopaedic implants and cardiovascular stents [1]. For instance, the fracture toughness of magnesium is greater than that of ceramic biomaterials and its elastic modulus and compressive yield strength are closer to those of natural bone [2,3]. Moreover, Mg is an essential metallic element in the human body with several functional roles and is present in the bone tissue; for instance, magnesium has a stimulatory effect on the growth of new bone tissue [4,5].

The main handicap of magnesium and its alloys is the low corrosion resistance that causes a reduction in the mechanical integrity of the implant. In order to overcome this limitation, different strategies have been reported and among them alloying elements and surface modifications stand out [6,7].

The alloying design strategy has been studied in depth in the case of magnesium alloys for structural applications and has been mainly focused on commercial Mg-Al-Zn (AZ) alloys [8] due to the beneficial effect of aluminium and zinc on castability, corrosion and mechanical properties [9,10]. On the contrary, the presence of elements such as Fe and Ni decrease the corrosion properties due to formation of micro-galvanic couples [11]. However, in those applications the toxicity and biocompatibility of the alloys are not taken into account [12]. In the biomedical filed, where biocompatibility is the main requirement, different elements have been studied following criteria of non-toxicity and absorbability in the human body [12–14].

Among them, calcium (Ca) and silver (Ag) present a great interest: (i) calcium is the main bone component, is an essential element in the human body and participates in cell signaling reactions [5,15]; (ii) silver is an element in the human body that has excellent antibacterial activity [16,17], it is also effective in the treatment of some microbes.

In the particular case of Mg alloys containing Ca and/or Ag, only a few works have been reported. For instance, the corrosion performance of MgXCa (X = 0.5–10 wt.% [18,19]) was studied and it was concluded that a low amount of Ca (up to 1%) decreases the corrosion rate, leads to bone regeneration around the implant and does not induce cytotoxicity. From the mechanical properties point of view, alloying with Ca (<4%) has been reported to increase the tensile strength [20,21].

Regarding magnesium alloys containing Ag, their antibacterial effect has been reported for a wide range of the microbial spectrum [22]. However, problems of high corrosion rates [16] and cytotoxicity [23] have also been found, although the cytotoxicity can be decreased while maintaining the antimicrobial effect if a secondary element, such as calcium, is included in the alloy [19].

Regarding the second strategy to improve the corrosion resistance based on surface modification of Mg and its alloys, the Plasma Electrolytic Oxidation (PEO) technique is one of the most promising candidates, as modified surfaces show improved corrosion behaviour along with other properties (e.g., biocompatibility) [24]. PEO generates anticorrosive, biocompatible and bioactive ceramic coatings with tailored composition, roughness, microstructure and porosity that can be controlled by optimization of the process electrical parameters and the composition of the electrolyte [24–26].

With respect to the PEO of Mg–Ca alloys, there are a few studies that reported an improved corrosion resistance [13,24,27,28], cell adhesion and bone regeneration [26,29,30]. However, to date there are no studies on PEO coatings on MgCaAg alloys.

The present work compares the corrosion resistance of Mg0.6Ca and Mg0.6Ca2Ag alloys in SBF with and without surface modification in order to determine their suitability for biodegradable implants.

2. Materials and Methods

Cast ingots of Mg0.6Ca and Mg0.6Ca2.0Ag (composition shown in Table 1) alloys were supplied by Magnesium Innovation Centre (MagIC, Helmholtz-Zentrum Geesthacht, Geesthacht, Germany). The ingots were cut into 60 × 6 × 4 mm^3 bars and ground on all sides to P1200 through successive grades of SiC abrasive paper, rinsed in isopropyl alcohol and dried in warm air. Electrical contact was provided at one end of the bars through a 1.5 mm metric threaded hole.

Table 1. Spark OES analysis of Mg0.6Ca and Mg0.6Ca2Ag alloy composition (wt.%).

Element	Mg0.6Ca	Mg0.6Ca2Ag	Element	Mg0.6Ca	Mg0.6Ca2Ag
Mg	99.4	97	Nd	0.00107	0.0009
Ca	>0.504	>0.504	Si	<0.0010	<0.0010
Ag	0.0005	2.417	Sr	0.0006	0.0007
Pr	0.02844	0.02579	Sn	<0.0005	<0.0005
Al	0.01985	<0.00020	Ni	0.0004	0.0005
Th	0.01547	0.01437	P	<0.0003	<0.0003
Mn	0.015	0.03715	Zr	<0.0003	<0.0003
Cu	0.00156	0.0004	Fe	<0.0002	<0.0002

PEO treatments were conducted using alternating current (AC) voltage-controlled EAC-S2000 (ET Systems electronic Gmbh, Altlußheim, Germany) power supply in a 2 L double jacket thermostated glass cell using a stainless steel mesh (AISI 316 of Ø15cm) as a counter electrode. The coatings were developed in a Ca–P containing electrolyte solution (10 g/L Na$_2$PO$_4$, 1 g/L KOH, 8 g/L NaF and 2.9 g/L

CaO) using a square voltage input waveform with positive and negative amplitudes of 430 and 50 V, respectively, 50% duty cycle, a frequency of 50 Hz and an rms current limitation of 138 mA/cm^2.

The alloy microstructure and coating composition and morphology were examined by scanning electron microscopy (JEOL JSM-6400, Tokyo, Japan) equipped with an energy dispersive X-ray (EDS) microanalysis system (OXFORD LINK PENTAFET 6506, Abingdon, Oxfordshire, UK). Metallographic preparation of the specimens included grinding from P120 to P1200 grit of SiC abrasive papers and polishing to 1 µm using diamond paste.

The alloys were also examined by transmission electron microscopy (TEM) using a JEM 2100HT JEOL instrument equipped with EDS operated at 200 keV of acceleration voltage. TEM specimens were prepared as 3 mm diameter and 0.1 µm-thick disks thinned until perforation by ion milling.

Surface potential maps of the alloys were obtained using a NANOSCOPE IIIA Multimode Scanning Kelvin Probe Force Microscope (SKPFM, Bruker, Billerica, MA, USA) with Pt coated Si tip operated in tapping mode. Topographic images and potential maps were obtained with the tip-to-sample distance fixed at 100 nm. The tip to sample distance was kept constant at 100 nm using a two-pass technique, where the height data is recorded in tapping mode during the first pass and the tip lifts above the surface to an adjustable lift height and scans the same line while following the height profile recorded in the second pass. All measurements were made at room temperature with a relative humidity in the range of 40%–65%.

X-ray diffraction (PHILIPS XPERT instrument, Amsterdam, The Netherlands) was used to characterize the phase components of the alloys and coatings using Cu-Kα = 1.54056 Å radiation and 2θ values between 10° and 90° at steps of 0.04°/2 s. A PANanalytical Aeris Research Edition was used to characterize Mg0.6ca2Ag using 2θ values between 10° and 30° at steps 0.04°/298.60 s. The spectra was analysed with the X-Pert High Score Plus program.

Hydrogen evolution measurements were carried out during immersion of PEO-coated and non-coated bars in m-SBF (5.403 g/L NaCl, 0.504 g/L NaHCO$_3$, 0.426 g/L Na$_2$CO$_3$, 0.225 g/L KCl, 0.230 g/L K$_2$HPO$_4$·3H$_2$O, 0.311 g/L MgCl$_2$·6H$_2$O, 17.892 g/L HEPES, 0.293 g/L CaCl$_2$, 0.072 g/L Na$_2$SO$_4$) [31] solution at 37 °C for up to 65 days using 20 mL/cm^2 of solution. The m-SBF was changed every 48 h or as needed if pH > 8.4. The electrical contact holes in the bars were sealed using a Lacquer 45 Stopping-off Resin (MacDermid Plc, Birmingham, UK).

Fluoride ions released during immersion of PEO-coated Mg0.6Ca alloy specimens (10.58 cm^2 of surface area) in 0.9% NaCl at 37 °C were measured using a fluoride ions selective electrode (Crison, Barcelona, Spain), which consisted of a lanthanum fluoride monocrystal membrane doped with europium. The membrane potential difference depended on the concentration of F$^-$ in the solution and was measured using a reference electrode of Ag/AgCl and pH meter GLP22 (Crison) every 30 min during first 2 h of immersion and at 24 h. The calibration curve was carried out using fluoride reference solutions (500 ppb, 1, 2.5, 5 and 10 ppm) prepared from a fluoride standard of 100 µg/L. Each reference solution was prepared using 25 mL of a TISAB solution (58 g/L NaCl, 4 g/L C$_{14}$H$_{22}$N$_2$O$_8$ in distilled water, pH adjusted to 5.0–5.5 with 6 M NaOH) in order to normalize the ionic strength.

3. Results

3.1. Characterisation

For both materials, dendritic α-Mg grains were revealed (Figure 1) with the second phase distributed along the grain boundaries in the form of a discontinuous network and decorated the interdendritic regions (bright contrast regions).

The EDS analysis of Areas 1 and 4 correspond to the matrix of Mg0.6Ca and Mg0.6Ca2Ag respectively (Table 2). Points 2 and 3 (Figure 1b) were located in the second phase of the Mg0.6Ca and revealed mainly a similar amount of Mg and Ca elements. In the case of the Mg0.6Ca2Ag alloy, two different EDS analyses (Figure 1d, Points 5 and 6) showed the formation of particles with different contents of Ag.

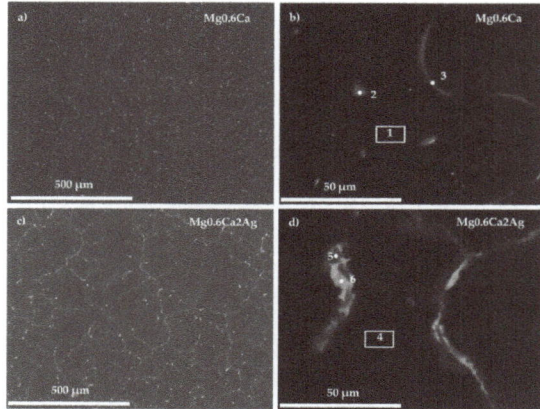

Figure 1. Backscattered electron micrographs of the studied alloys: (**a**,**b**) Mg0.6Ca and (**c**,**d**) Mg0.6Ca2Ag.

Table 2. Local EDS point analysis as per Figure 1.

Alloy	Location	Mg	Ca	Al	Si	Ag
Mg0.6Ca	1	96.77	2.3	0.73	0.2	–
	2	90.2	9.42	0.38	–	–
	3	92.66	7.11	0.27	–	–
Mg0.6Ca2Ag	4	99.4	0.15	–	–	0.4
	5	98.13	0.3	–	–	1.57
	6	76.03	7.47	–	–	15.57

TEM analyses of Mg0.6Ca alloy displayed the second phase (Figure 2a) as an eutectic aggregate with lamellar morphology, probably formed by α-Mg/Mg$_2$Ca phases in accordance with the EDS analysis (Table 3), the Mg–Ca phase diagram and other works on binary Mg–Ca alloys [25,32–34]. In addition, polygonal shape particles (Figure 2b) containing impurities were found (Point 2, Table 3).

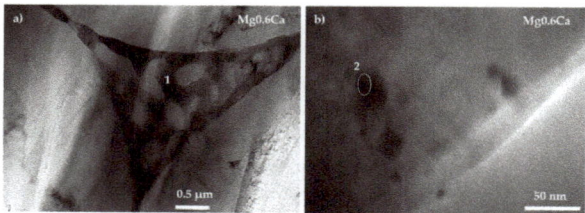

Figure 2. Transmission electron micrographs of the Mg0.6Ca alloy with locations of EDS analysis: (**a**) Mg0.6Ca 0.5 µm and (**b**) Mg0.6Ca 0.5 nm.

Table 3. Local EDS analysis (at.%) of intermetallic particles as per Figures 2 and 3.

Alloy	Location	Mg	Ca	Ag	Fe	Al	Ni	Co	Cr	Si
Mg0.6Ca	1	92.59	7.08	–	–	0.33	–	–	–	–
	2	85.04	12.7	–	0.05	1.29	0.86	–	–	–
Mg0.6Ca2Ag	3	74.6	7.27	16.52	0.59	–	0.15	0.51	0.62	0.15
	4	99.08	0.15	0.31	0.15	–	–	0.12	0.18	–

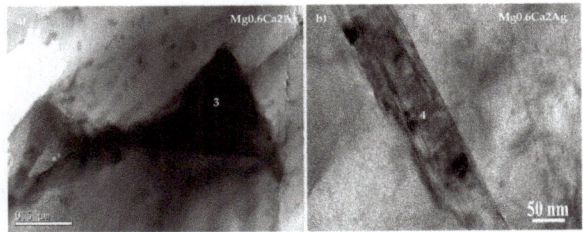

Figure 3. Transmission electron micrographs of the Mg0.6Ca2Ag alloy with locations of EDS analysis. (**a**) Mg0.6Ca2Ag 0.5 µm and (**b**) Mg0.6Ca2Ag 50 nm.

In the case of the ternary alloy (Mg0.6Ca2Ag) (Figure 3), TEM studies were not conclusive as the zones of the ion milled lamella where particles were located were too thick for generating an informative electron diffraction pattern.

In fact, the available data of TEM studies on Mg alloys containing silver are very scarce and all of them have been conducted on binary Mg–Ag alloys reporting the formation of different intermetallic compounds such as MgAg, $MgAg_3$ or $MgAg_4$ [35,36]. The present work revealed the formation of particles with different shapes (Figure 3) and compositions (Table 4), some of them were Ag-enriched (Table 3, Point 2).

Table 4. EDS analysis (at.%) of Mg0.6Ca alloys at the locations of SKPFM mapping as per Figure 4.

Alloy	Location	Mg	Ca	Si	O	Al
	1	91.31	6.15	–	2.23	0.3
Mg0.6Ca	2	80.42	7.38	10.87	1.33	–
	3	95.67	4.33	–	–	–

The local Volta potential difference (VPD) between constituents on a submicron scale was obtained using SKPFM. Surface potential maps and potential profile along with the SEM images of the studied area are displayed in Figure 4 for the Mg0.6Ca and Mg0.6Ca2Ag alloys, respectively.

Figure 4a,d show the selected region of the Mg0.6Ca alloy where surface potential maps and profiles were acquired. Point 1 corresponds to the Mg_2Ca phase that shows a negative potential with respect to the α-Mg matrix (VPD of ~−50 mV). It is important to note that the available data regarding the electrochemical behaviour of Mg_2Ca is quite limited and controversial results are reported about the anodic [37,38] as well as cathodic behaviour [19] of this phase, which have been mainly attributed to the differences in the composition of the alloy (e.g., presence of impurities) and matrix segregation. Points 2 and 3 (Figure 4d) correspond to an impurity (Point 2) and to the Mg_2Ca phase (Point 3, Table 4). As it was observed before, the latter shows an anodic behaviour (VPD ~−15 mV) compared to the matrix, whereas the presence of impurities reveals a slight cathodic behaviour with respect to the α-Mg matrix (VPD ~+20 mV) due to the presence of more noble elements such as Si, Fe or Al.

For the Mg0.6Ca2Ag alloy, the Volta potential profiles conducted in the inclusion (Figure 5) reveal a different electrochemical response depending on the elemental composition analyzed by EDS (Table 5). The area enriched in Ag (Ca/Ag ratio of 0.89) (Figure 5b) revealed a cathodic behaviour (VPD ~+20 mV) in comparison with the α-Mg matrix, whereas the Ca/Ag ratio of 2.39 leads to an anodic performance with potential differences around VPD ~−27 mV. Ben et al. studied the electrochemical behaviour of magnesium–silver ternary alloys (MgZnAg) and reported a cathodic behaviour of MgAg secondary phase [39], which appears to be similar to the behaviour of the areas enriched with Ag in Mg0.6Ca2Ag alloy of the present work.

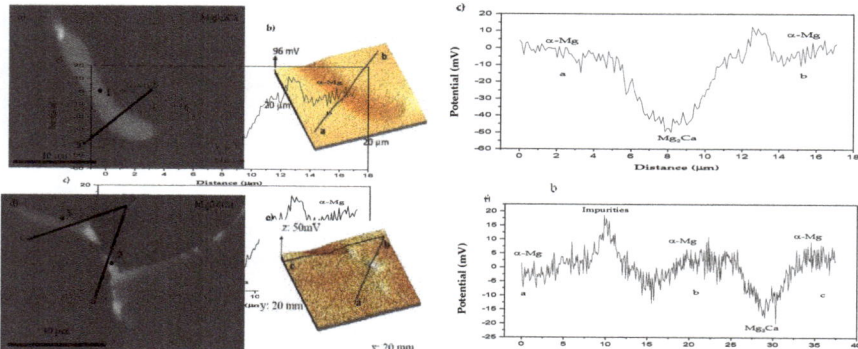

Figure 4. SEM backscattered images (**a,d**), surface potential maps (**b,e**) and potential profile (**c,f**) in selected areas of the Mg0.6Ca alloy.

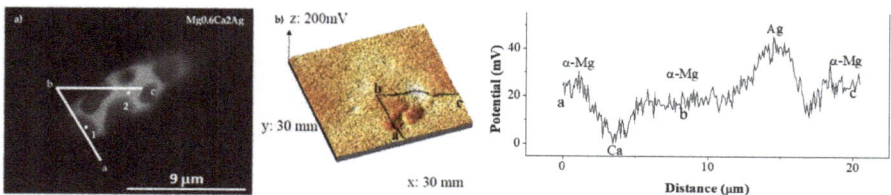

Figure 5. SEM backscattered image (**a**), surface potential maps (**b**) and potential profile (**c**) in selected areas of the Mg0.6Ca2Ag alloy.

Table 5. EDS analysis (at.%) of the Mg0.6Ca2Ag alloy at the locations of SKPFM mapping as per Figure 5.

Alloy	Location	Mg	Ca	O	Ag
Mg0.6Ca2Ag	1	91.44	5.04	1.42	2.11
	2	81.67	8.32	0.71	9.31

Figure 6 shows the plan view and cross-sectional coating morphologies of Mg0.6Ca/PEO and Mg0.6Ca2Ag/PEO coatings. In both alloys the coating surface presents a typical crater-like porous morphology associated with the sites of discharge channels, gas evolution and rapid solidification phenomena [24,40]. In both Mg0.6Ca/PEO and Mg0.6Ca2Ag/PEO coatings the surface Ca/P ratio is relatively high (1.42 and 1.53, respectively) compared with the inner regions of the coating (Table 6), although lower than that of biological hydroxyapatite (1.67) [41], and both Ca and P contents increase towards the coating/electrolyte interface. The surface enrichment of biomaterials (e.g., of Ti alloys) in Ca and P is well known to improve the initial cell response [42].

The cross-sectional images (Figure 6c,d) reveal relatively uniform coatings with thicknesses in the range of 32–35 μm. It can be observed that both coatings are constituted by three layers. A thin barrier layer (less than 1 μm) adjacent to the substrate is mainly composed of Mg, O and F and Ag (the latter only in the case of MgCaAg, Table 6). An intermediate region with small pores constitutes ~40% of the coating thickness. An outer, more compact, region contains a few but relatively large pores.

It is worth mentioning that EDS analysis for the MgCaAg alloy did not detect the presence of Ag in the coating surface, whereas the Ag content in the inner regions did not exceed ~0.2 at.%, suggesting that a rather limiting if any antibacterial effect can be expected at the initial stage of the implantation.

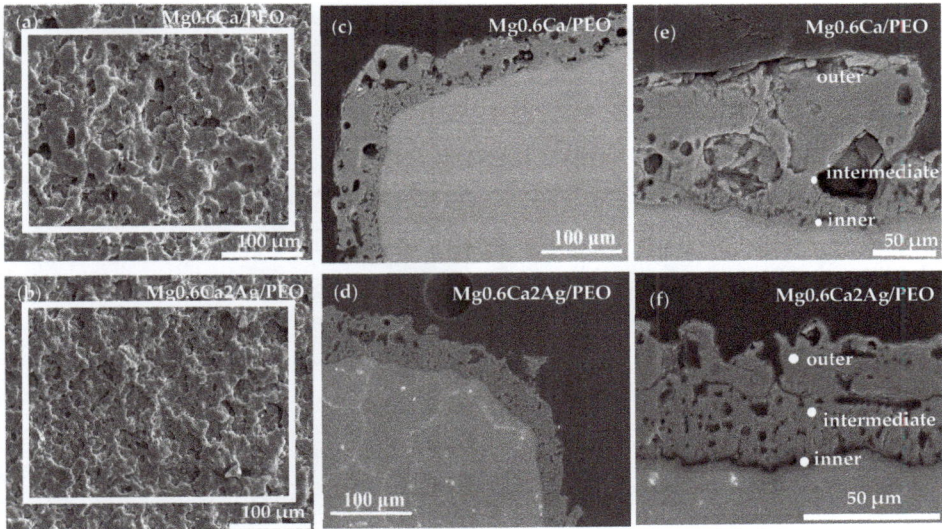

Figure 6. Backscattered electron micrographs of coating surface morphologies and cross-sections after corrosion of Mg0.6Ca/PEO (**a,c,e**) and PEO-Mg0.6Ca2Ag (**b,d,f**).

Table 6. Compositions of Mg0.6Ca/PEO and Mg0.6Ca2Ag/PEO before corrosion by EDX (at.%).

Alloy	Location	Mg	O	F	Na	P	Ca	Ag	Ca/P
Mg0.6Ca	Surface	22.1	44.9	16.3	4.2	5.2	7.3	–	1.42
	Inner layer	39.7	24.5	32.5	0.3	2.9	0.1	–	0.63
	Intermediate layer	35.2	21.2	36.7	1.9	4.4	0.6	–	0.14
	Outer layer	36.5	34.9	18.6	4.2	3.6	2.3	–	0.05
Mg0.6Ca2Ag	Surface	27	42.5	16.2	3.7	4.2	6.4	–	1.53
	Inner layer	37.6	22.3	34.7	1	3.8	0.5	0.2	0.4
	Intermediate layer	32.4	26.3	30.1	4.3	5.1	1.5	0.2	0.3
	Outer layer	36.2	37.3	13	4.6	6.3	2.5	–	0.12

The XRD analyses of uncoated materials revealed peaks corresponding to α-Mg for both alloys (Figure 7). In the case of uncoated Mg0.6Ca, no intermetallic/secondary phases were detected probably due to their negligible amount. Although, in our previous study of a similar alloy (cast Mg0.8Ca) a formation of Mg_2Ca phase was confirmed [25]. Most of the diffraction peaks of Mg0.6Ca2Ag were the same as those for Mg0.6Ca except that there were small peaks in a low 2θ angles region corresponding to binary and ternary intermetallic phases as Mg_2Ca, $MgAg_4$, $Mg_{54}Ag_{17}$ [25,36,43] and Ca_2MgAg_3 (Figure 7, inset). A further systematic TEM and electron diffraction study would be necessary in order to confirm the presence of these phases.

On the other hand, PEO coatings revealed peaks corresponding to the substrate, and the formation of crystalline phases such as MgO, MgF_2, CaF_2 and $Ca_5(PO_4)3F$ was detected. The MgO is formed due to electrolytic oxidation of the substrate and the other phases are formed due to plasma-chemical reactions between the ions of the electrolyte and the substrate inside the microdischarge channels [44].

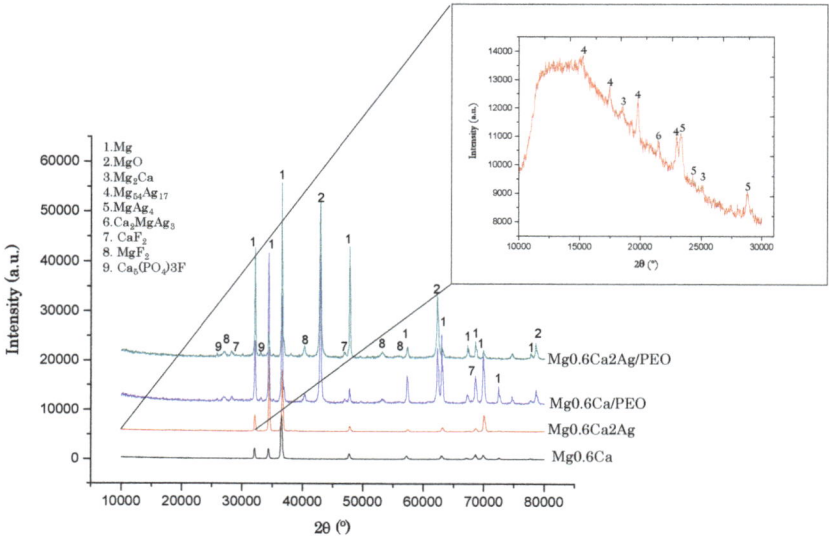

Figure 7. XDR patterns from bulk material and PEO coatings.

3.2. Hydrogen Evolution Measurement

Figure 8a,b show the hydrogen evolution volume and hydrogen evolution rate for Mg0.6Ca and Mg0.6Ca/PEO after 60 days of immersion in SBF at 37 °C. Figure 9 shows the volume of evolved hydrogen for Mg0.6Ca2Ag and Mg0.6Ca2Ag/PEO after 4 days of immersion. As expected, the uncoated material initially exhibited a high amount of hydrogen with an evolution rate of 3.86 mL/cm^2 week after 1 week of immersion, with progressive decrease to 1.02 and 1.11 mL/cm^2 week after 6 and 8 weeks of immersion, respectively. The decrease of hydrogen evolution rate is due to the generation of a corrosion products layer, which acts as a partially protective coating.

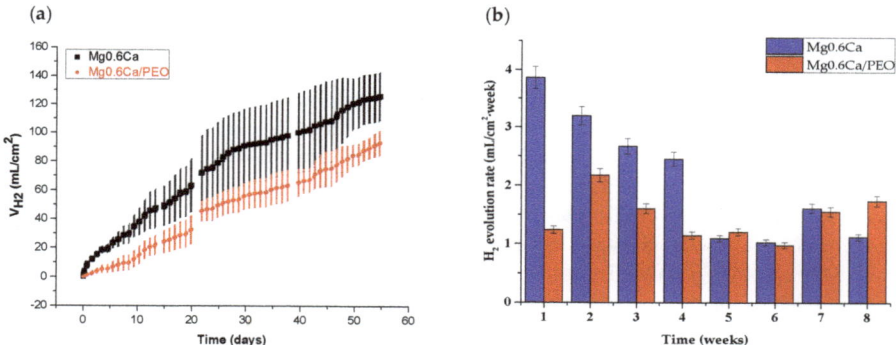

Figure 8. (a) Hydrogen volume and (b) hydrogen evolution rate of Mg0.6Ca and Mg0.6Ca/PEO after 60 days of immersion in m-SBF.

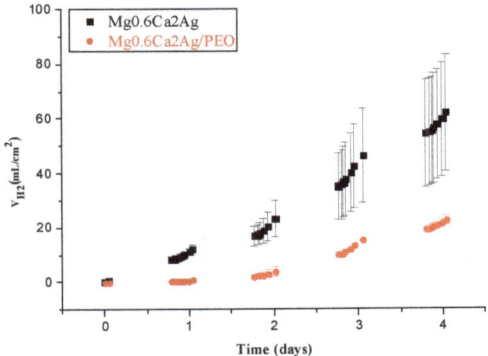

Figure 9. Hydrogen volume for Mg0.6Ca2Ag and Mg0.6Ca2Ag/PEO after 4 days of immersion in m-SBF.

Mg0.6Ca/PEO showed a considerably reduced hydrogen evolution rate during the first 4 weeks (~1.11 mL/cm^2 week); however, after that time both non-coated and PEO-coated materials reached a similar degradation rate. Further increase of the hydrogen evolution rate was evident after 6 weeks (1.75 mL/cm^2 week) due to the loss of protective properties of the inner PEO layer and increased electrochemical activities in the substrate/coating interface [45–47]. The latter degradation rate corresponds to 1.9 mg/(cm^2 week) of mass loss or 11 μm/week of thickness loss.

Both Mg0.6Ca2Ag and Mg0.6Ca2Ag/PEO exhibited an extremely high degradation rate (Figure 9), corresponding to ~60 or ~20 mL/cm^2, respectively, and the experiments were stopped after 4 days of immersion.

Figure 10 shows the macro degradation photos of the specimens. Both coated and non-coated Mg0.6Ca alloy (Figure 10a,b) presented a uniform corrosion with similar loss of material and dimensions. Mg0.6Ca2Ag also showed a generalized, but heavily heterogeneous corrosion morphology for the non-coated specimens (Figure 10c) and localized corrosion for the PEO-coated ones; both were found completely disintegrated after about 10 days of immersion.

 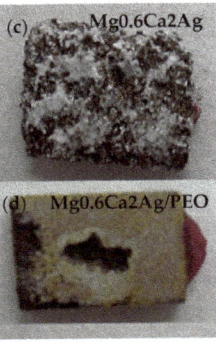

Figure 10. 3D macrograph of (**a**) Mg0.6Ca, (**b**) Mg0.6Ca/PEO after 60 days, (**c**) Mg0.6Ca2Ag and (**d**) Mg0.6Ca2Ag/PEO after 4 days of immersion in m-SBF.

Following the immersion, a heavy generalized corrosion was observed in all cases (Figure 11) and PEO coatings were evidently detached, and, in the case of the Mg0.6Ca2Ag/PEO system, this had already occurred by the 4th day of immersion. A complete loss of adhesion is not a typical behaviour for PEO-coated Mg alloy, as was previously demonstrated by the authors in [25], and a 40–50 μm-thick coating can be expected to remain mostly adhered even after 8 weeks of immersion with a thick corrosion product layer developing underneath it.

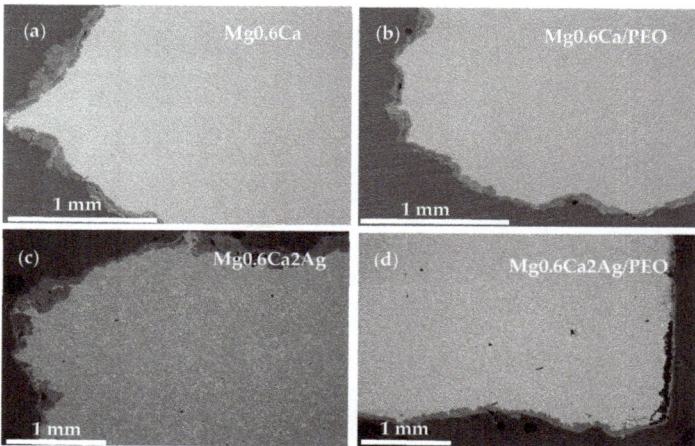

Figure 11. Cross-section before corrosion (**a**) Mg0.6Ca, (**b**) Mg0.6Ca/PEO after 60 days and (**c**) Mg0.6Ca2Ag and (**d**) Mg0.6Ca2Ag/PEO after 4 days of immersion in SBF.

Some researchers of Mg–Ca systems have reported that for Ca content below 1.25% the corrosion process is driven from a general mechanism to localized corrosion due to severe electrochemical activity [18]. Other studies suggest that when the amount of Ag in MgAg alloys is increased [16] (1.15 mm/year for Mg2Ag to 1.43 mm/year for Mg6Ag) the Ag-containing second phase does not play an important role in microgalvanic activities, as the responsibility for these phenomena fall to the impurities [48,49]. However, in our case the corrosion rate was 7.78 mm/year for Mg0.6Ca2Ag and 3.64 mm/year for Mg0.6Ca2Ag/PEO. The elevated degradation rate was observed in the present study for Mg0.6Ca in comparison with other Mg–Ca systems [25] and especially for Mg0.6Ca2Ag.

It is evident that, in both alloys, intermetallic particles disclosed some regions with active behaviour that depended on the Ca/Ag ratio in the region (Figures 4 and 5). Such regions present galvanic micro-couples with a minimum anodic surface compared to a large cathodic surface of α-Mg causing dissolution of the intermetallic particles. It can be clearly appreciated from Figure 1a,c that in the Mg0.6Ca2Ag alloy the grain boundary network is much more decorated with intermetallic particles than in Mg0.6Ca. Consequently, dissolution of the grain boundaries can lead to an easy fall-out and loss of the cathodic α-Mg grains or clusters of grains (Figure 12). This anodic behavior of the particles may explain the extremely fast degradation rate of the Mg0.6Ca2Ag alloy.

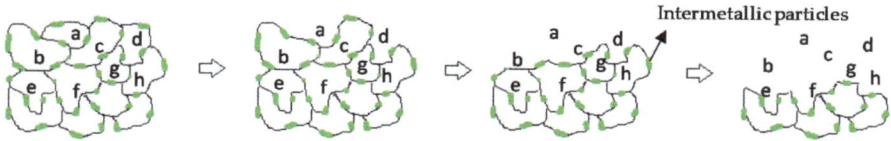

Figure 12. Schematic diagram of corrosion attack at the grain boundary and loss of grains (labeled with letters).

3.3. Fluoride Release

Figure 13 shows progressive F$^-$ ion release from Mg0.6Ca/PEO during 24 h of immersion in 0.9% NaCl. After 24 h of immersion, 33.757 µg/cm^2 of F$^-$ was released, which corresponds to 1.35 ppm or 71 µM of F⁻ for the volume of the immersion solution. According to the solubility products, K_{sp}, of fluoride containing crystalline phases in the coating, MgF_2, CaF_2, and $Ca_5(PO_4)_3F$ (5.16×10^{-11}, 3.58×10^{-11}, and 8.6×10^{-61}, respectively [50]), the dissolution of these compounds will produce up to a maximum of 0.88 mM of free F⁻. Hence, it would be reasonable to expect a further increase of F⁻ with time until a solubility limit is reached. In our previous work, it was demonstrated that this limit can be reached in 12 weeks [33]. For comparison, according to The World Health Organization (WHO), fluoride content in drinking water should not exceed 1.22 mg/L (~1 ppm), the optimal daily fluoride consumption for an adult is between 1.4–3.4 mg·day⁻¹, whereas the average fluorine concentration in the human blood plasma is 19 ppb [51]. Therefore, the fluoride liberation from Mg0.6Ca/PEO in a 24-h period appears to be safely within these guidelines without a danger of intoxication. Further studies are needed in order to evaluate the potential antibacterial effect of the fluoride in the coating.

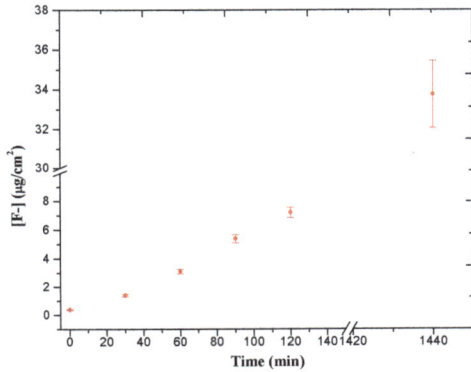

Figure 13. Fluoride ions released of PEO coating for 120 min.

4. Conclusions

- PEO coatings of 32–35 µm thickness, containing bioactive calcium fluoride, magnesium fluoride and fluorapatite phases, were generated on Mg0.6Ca and Mg0.6Ca2Ag alloys for biomedical implant applications.
- The hydrogen evolution rate of PEO-coated Mg0.6Ca 3D prototypes evaluated during long-term immersion tests (up to 60 days), was 3.86 mL/cm² week during the first month of immersion in SBF and 1.75 mL/cm² week during the second month of immersion. The uncoated alloy generated two times more hydrogen in the first month of the test.
- The degradation mechanism during immersion of the studied materials corresponds to a generalized corrosion in both alloys. The Mg0.6Ca2Ag alloys degrade much faster than Mg0.6Ca due to the varied content of Ag and impurities in Mg–Ca–Ag intermetallic particles, which are mainly distributed along the grain boundaries; some of them acting as local anodes and causing the loss of the whole grain or clusters of grains. This also happens in Mg0.6Ca but to a much lesser extent, which leads to a slower corrosion process.
- The PEO coatings liberate fluoride ions during immersion in 0.9 M NaCl. The fluoride release is increased throughout 24 h of immersion time for Mg0.6Ca/PEO. The fluoride reserves of the coating have not been completely depleted in the course of the 24 h immersion.

Author Contributions: Formal Analysis, L.M., M.M., B.M., R.A. and E.M.; Funding Acquisition, R.A. and E.M.; Investigation and Methodology L.M. and M.M.; Resources, M.M. and E.M.; Supervision, E.M. and R.A.; Writing—Original Draft Preparation, L.M. and B.M.; Writing—Review & Editing, E.M. and R.A.; Conceptualization, L.M., M.M., B.M., R.A. and E.M.; Software, L.M., M.M., B.M., R.A. and E.M.; Validation, L.M., M.M., B.M., R.A. and E.M.; Data Curation, L.M., M.M. and B.M.; Visualization, L.M., M.M., B.M., R.A. and E.M.; Project Administration, R.A. and E.M.

Funding: This work was partially supported by (MAT2015-73355-JIN) and ADITIMAT-CM (S2018/NMT-4411). M.M. is grateful the Ramon y Cajal Programme (MICINN, Spain, RYC-2017-21843).

Acknowledgments: The authors would like to acknowledge the Magnesium Innovation Centre (MagIC, Helmholtz-Zentrum Geesthacht) for supplying the alloys.

Conflicts of Interest: The authors declare no conflict of interest.

References

1. Sanchez, A.H.M.; Luthringer, B.J.; Feyerabend, F.; Willumeit, R. Mg and Mg alloys: How comparable are in vitro and in vivo corrosion rates? A review. *Acta Biomater.* **2014**, *13*, 16–31. [CrossRef] [PubMed]
2. Ding, W. Opportunities and challenges for the biodegradable magnesium alloys as next-generation biomaterials. *Regen. Biomater.* **2016**, *3*, 79–86. [CrossRef] [PubMed]
3. Antoniac, I.; Laptoiu, D. Magnesium alloys—Current orthopedic applications. *Revista de Ortopedie si Traumatologie a Asociatiei de Ortopedie Româno-Italo-Spaniole* **2010**, *4*, 79–88.
4. Manivasagam, G.; Suwas, S. Biodegradable Mg and Mg based alloys for biomedical implants. *Mater. Sci. Technol.* **2014**, *30*, 515–520. [CrossRef]
5. Staiger, M.P.; Pietak, A.M.; Huadmai, J.; Dias, G. Magnesium and its alloys as orthopedic biomaterials: A review. *Biomaterials* **2006**, *27*, 1728–1734. [CrossRef]
6. Waizy, H.; Seitz, J.M.; Reifenrath, J.; Weizbauer, A.; Bach, F.W.; Meyer-Lindenberg, A.; Denkena, B.; Windhagen, H. Biodegradable magnesium implants for orthopedic applications. *J. Mater. Sci.* **2013**, *48*, 39–50. [CrossRef]
7. Walker, J.; Shadanbaz, S.; Woodfield, T.B.F.; Staiger, M.P.; Dias, G.J. Magnesium biomaterials for orthopedic application: A review from a biological perspective. *J. Biomed. Mater. Res. B* **2014**, *102*, 1316–1331. [CrossRef]
8. Feliu, S., Jr.; Pardo, A.; Merino, M.C.; Coy, A.E.; Viejo, F.; Arrabal, R. Correlation between the surface chemistry and the atmospheric corrosion of AZ31, AZ80 and AZ91D magnesium alloys. *Appl. Surf. Sci.* **2009**, *255*, 4102–4108. [CrossRef]
9. Esmaily, M.; Svensson, J.E.; Fajardo, S.; Birbilis, N.; Frankel, G.S.; Virtanen, S.; Arrabal, R.; Thomas, S.; Johansson, L.G. Fundamentals and advances in magnesium alloy corrosion. *Prog. Mater. Sci.* **2017**, *89*, 92–193. [CrossRef]
10. Pardo, A.; Merino, M.C.; Coy, A.E.; Viejo, F.; Arrabal, R.; Feliú, S., Jr. Influence of microstructure and composition on the corrosion behaviour of Mg/Al alloys in chloride media. *Electrochim. Acta* **2008**, *53*, 7890–7902. [CrossRef]
11. Makar, G.L.; Kruger, J. Corrosion of magnesium. *Int. Mater. Rev.* **1993**, *38*, 138–153. [CrossRef]
12. Chen, Y.; Xu, Z.; Smith, C.; Sankar, J. Recent advances on the development of magnesium alloys for biodegradable implants. *Acta Biomater.* **2014**, *10*, 4561–4573. [CrossRef]
13. Radha, R.; Sreekanth, D. Insight of magnesium alloys and composites for orthopedic implant applications—A review. *J. Magnes. Alloy.* **2017**, *5*, 286–312. [CrossRef]
14. Zhang, C.; Lin, J.; Liu, H. Magnesium-based biodegradable materials for biomedical applications. *MRS Adv.* **2018**, *3*, 2359–2364. [CrossRef]
15. Gusieva, K.; Davies, C.H.J.; Scully, J.R.; Birbilis, N. Corrosion of magnesium alloys: The role of alloying. *Int. Mater. Rev.* **2015**, *60*, 169–194. [CrossRef]
16. Tie, D.; Feyerabend, F.; Müller, W.-D.; Schade, R.; Liefeith, K.; Kainer, K.; Willumeit, R. Antibacterial biodegradable Mg-Ag alloys. *Eur. Cells Mater.* **2013**, *25*, 284–298. [CrossRef]
17. Chen, Q.; Thouas, G.A. Metallic implant biomaterials. *Mater. Sci. Eng. R.* **2015**, *87*, 1–57. [CrossRef]
18. Rad, H.R.B.; Idris, M.H.; Kadir, M.R.A.; Farahany, S. Microstructure analysis and corrosion behavior of biodegradable Mg-Ca implant alloys. *Mater. Des.* **2012**, *33*, 88–97. [CrossRef]
19. Li, Z.; Gu, X.; Lou, S.; Zheng, Y. The development of binary Mg-Ca alloys for use as biodegradable materials within bone. *Biomaterials* **2008**, *29*, 1329–1344. [CrossRef]
20. Salahshoor, M.; Guo, Y. Biodegradable orthopedic magnesium-calcium (MgCa) alloys, processing, and corrosion performance. *Materials* **2012**, *5*, 135–155. [CrossRef]
21. Mohedano, M.; Arrabal, R.; Mingo, B.; Pardo, A.; Matykina, E. Role of particle type and concentration on characteristics of PEO coatings on AM50 magnesium alloy. *Surf. Coat. Technol.* **2018**, *334*, 328–335. [CrossRef]

22. Yu, K.; Dai, Y.; Luo, Z.; Long, H.; Zeng, M.; Li, Z.; Zhu, J.; Cheng, L.; Zhang, Y.; Liu, H.; et al. In vitro and in vivo evaluation of novel biodegradable Mg-Ag-Y alloys for use as resorbable bone fixation implant. *J. Biomed. Mater. Res. A* **2018**, *106*, 2059–2069. [CrossRef] [PubMed]
23. Gopi, D.; Shinyjoy, E.; Kavitha, L. Synthesis and spectral characterization of silver/magnesium co-substituted hydroxyapatite for biomedical applications. *Spectrochim. Acta Part A* **2014**, *127*, 286–291. [CrossRef] [PubMed]
24. Mohedano, M.; Lu, X.; Matykina, E.; Blawert, C.; Arrabal, R.; Zheludkevich, M. Plasma electrolytic oxidation (PEO) of metals and alloys. *Encycl. Interfacial Chem.* **2018**, 423–438.
25. Mohedano, M.; Luthringer, B.J.C.; Mingo, B.; Feyerabend, F.; Arrabal, R.; Sanchez-Egido, P.J.; Blawert, C.; Willumeit-Römer, R.; Zheludkevich, M.L.; Matykina, E. Bioactive plasma electrolytic oxidation coatings on Mg-Ca alloy to control degradation behavior. *Surf. Coat. Technol.* **2017**, *315*, 454–467. [CrossRef]
26. Gu, X.; Li, N.; Zhou, W.; Zheng, Y.; Zhao, X.; Cai, Q.; Ruan, L. Corrosion resistance and surface biocompatibility of a microarc oxidation coating on a Mg-Ca alloy. *Acta Biomater.* **2011**, *7*, 1880–1889. [CrossRef]
27. Jugdaohsingh, R. Silicon and bone health. *J. Nutr. Health Aging* **2007**, *11*, 99–110.
28. Pan, Y.; Chen, C.; Feng, R.; Cui, H.; Gong, B.; Zheng, T.; Ji, Y. Effect of calcium on the microstructure and corrosion behavior of microarc oxidized Mg-xCa alloys. *Biointerphases* **2018**, *13*, 011003. [CrossRef]
29. Narayanan, T.S.; Park, I.S.; Lee, M.H. Strategies to improve the corrosion resistance of microarc oxidation (MAO) coated magnesium alloys for degradable implants: Prospects and challenges. *Prog. Mater. Sci.* **2014**, *60*, 1–71. [CrossRef]
30. Kim, E.-J.; Bu, S.-Y.; Sung, M.-K.; Choi, M.-K. Effects of silicon on osteoblast activity and bone mineralization of MC3T3-E1 cells. *Biol. Trace Elem. Res.* **2013**, *152*, 105–112. [CrossRef]
31. Salahshoor, M.; Guo, Y.B. Biodegradation control of magnesium-calcium biomaterial via adjusting surface integrity by synergistic cutting-burnishing. *Procedia CIRP* **2014**, *13*, 143–149. [CrossRef]
32. Kirkland, N.T.; Birbilis, N.; Walker, J.; Woodfield, T.; Dias, G.J.; Staiger, M.P. In-vitro dissolution of magnesium-calcium binary alloys: Clarifying the unique role of calcium additions in bioresorbable magnesium implant alloys. *J. Biomed. Mater. Res. Part B* **2010**, *95*, 91–100. [CrossRef]
33. Bita, A.I.; Antoniac, A.; Cotrut, C.; Vasile, E.; Ciuca, I.; Niculescu, M.; Antoniac, I. In vitro degradation and corrosion evaluation of Mg-Ca alloys for biomedical applications. *J. Optoelectron. Adv. Mater.* **2016**, *18*, 394–398.
34. Deng, M.; Höche, D.; Lamaka, S.V.; Snihirova, D.; Zheludkevich, M.L. Mg-Ca binary alloys as anodes for primary Mg-air batteries. *J. Power Sources* **2018**, *396*, 109–118. [CrossRef]
35. Wang, H.; Lou, Y.; Northwood, D.O. Synthesis of Ag nanoparticles by hydrolysis of Mg-Ag intermetallic compounds. *J. Mater. Process. Technol.* **2008**, *204*, 327–330. [CrossRef]
36. Du, J.; Zhang, A.; Guo, Z.; Yang, M.; Li, M.; Xiong, S. Atomic cluster structures, phase stability and physicochemical properties of binary Mg-X (X = Ag, Al, Ba, Ca, Gd, Sn, Y and Zn) alloys from ab-initio calculations. *Intermetallics* **2018**, *95*, 119–129. [CrossRef]
37. Bakhsheshi-Rad, H.; Abdul-Kadir, M.; Idris, M.; Farahany, S. Relationship between the corrosion behavior and the thermal characteristics and microstructure of Mg-0.5Ca-xZn alloys. *Corros. Sci.* **2012**, *64*, 184–197. [CrossRef]
38. Yang, J.; Peng, J.; Nyberg, E.A.; Pan, F.-S. Effect of Ca addition on the corrosion behavior of Mg-Al-Mn alloy. *Appl. Surf. Sci.* **2016**, *369*, 92–100. [CrossRef]
39. Ben-Hamu, G.; Eliezer, D.; Kaya, A.; Na, Y.; Shin, K. Microstructure and corrosion behavior of Mg-Zn-Ag alloys. *Mater. Sci. Eng. A* **2006**, *435–436*, 579–587. [CrossRef]
40. Yerokhin, A.; Nie, X.; Leyland, A.; Matthews, A.; Dowey, S. Plasma electrolysis for surface engineering. *Surf. Coat. Technol.* **1999**, *122*, 73–93. [CrossRef]
41. Scapin, M.A.; Guilhen, S.N.; Cotrim, M.E.; Pires, M.A.F. Determination of Ca/P molar ratio in hydroxyapatite (HA) by X-ray fluorescence technique. Proceedings of International Nuclear Atlantic Conference Brazilian Nuclear Program State Policy for A Sustainable World, Sao Paulo, Brazil, 4–9 October 2015.
42. Mohedano, M.; Matykina, E.; Arrabal, R.; Pardo, A.; Merino, M. Metal release from ceramic coatings for dental implants. *Dent. Mater.* **2014**, *30*, 28–40. [CrossRef] [PubMed]
43. Laws, K.J.; Shamlaye, K.F.; Granata, D.; Koloadin, L.S.; Löffler, J.F. Electron-band theory inspired design of magnesium-precious metal bulk metallic glasses with high thermal stability and extended ductility. *Sci. Rep.* **2017**, *7*, 3400. [CrossRef] [PubMed]

44. Matykina, E.; Garcia, I.; Arrabal, R.; Mohedano, M.; Mingo, B.; Sancho, J.; Merino, M.; Pardo, A. Role of PEO coatings in long-term biodegradation of a Mg alloy. *Appl. Surf. Sci.* **2016**, *389*, 810–823. [CrossRef]
45. Gao, Y.; Yerokhin, A.; Matthews, A. DC plasma electrolytic oxidation of biodegradable cp-Mg: In-vitro corrosion studies. *Surf. Coat. Technol.* **2013**, *234*, 132–142. [CrossRef]
46. Arrabal, R.; Matykina, E.; Viejo, F.; Skeldon, P.; Thompson, G. Corrosion resistance of WE43 and AZ91D magnesium alloys with phosphate PEO coatings. *Corros. Sci.* **2008**, *50*, 1744–1752. [CrossRef]
47. Gao, Y.; Yerokhin, A.; Matthews, A. Deposition and evaluation of duplex hydroxyapatite and plasma electrolytic oxidation coatings on magnesium. *Surf. Coat. Technol.* **2015**, *269*, 170–182. [CrossRef]
48. Mathieu, S.; Rapin, C.; Steinmetz, J.; Steinmetz, P. A corrosion study of the main constituent phases of AZ91 magnesium alloys. *Corros. Sci.* **2003**, *45*, 2741–2755. [CrossRef]
49. Ballerini, G.; Bardi, U.; Bignucolo, R.; Ceraolo, G. About some corrosion mechanisms of AZ91D magnesium alloy. *Corros. Sci.* **2005**, *47*, 2173–2184. [CrossRef]
50. McCann, H. The solubility of fluorapatite and its relationship to that of calcium fluoride. *Arch. Oral Biol.* **1968**, *13*, 987–1001. [CrossRef]
51. Agalakova, N.I.; Gusev, G.P. Molecular mechanisms of cytotoxicity and apoptosis induced by inorganic fluoride. *ISRN Cell Biol.* **2012**, *2012*, 403835. [CrossRef]

© 2019 by the authors. Licensee MDPI, Basel, Switzerland. This article is an open access article distributed under the terms and conditions of the Creative Commons Attribution (CC BY) license (http://creativecommons.org/licenses/by/4.0/).

Article

Influence of SiO₂ Particles on the Corrosion and Wear Resistance of Plasma Electrolytic Oxidation-Coated AM50 Mg Alloy

Xiaopeng Lu [1,2,3,*], Yan Chen [1,2,*], Carsten Blawert [3], Yan Li [1,2], Tao Zhang [1,2], Fuhui Wang [1,2], Karl Ulrich Kainer [3] and Mikhail Zheludkevich [3,4]

1. Corrosion and Protection Division, Shenyang National Laboratory for Materials Science, Northeastern University, Shenyang 110819, China; liyan93@stumail.neu.edu.cn (Y.L.); zhangtao@mail.neu.edu.cn (T.Z.); fhwang@mail.neu.edu.cn (F.W.)
2. Key Laboratory for Anisotropy and Texture of Materials (Education Ministry of China), Northeastern University, Shenyang 110004, China
3. Magnesium Innovation Centre (MagIC), Helmholtz-Zentrum Geesthacht, Max-Planck-Str. 1, 21502 Geesthacht, Germany; carsten.blawert@hzg.de (C.B.); karl.kainer@hzg.de (K.U.K); mikhail.zheludkevich@hzg.de (M.Z.)
4. Faculty of Engineering, University of Kiel, Kaiserstrasse 2, 24143 Kiel, Germany
* Correspondence: luxiaopeng@mail.neu.edu.cn (X.L.); chenyan2501@hotmail.com (Y.C.)

Received: 10 July 2018; Accepted: 20 August 2018; Published: 29 August 2018

Abstract: The influence of SiO₂ particles on the microstructure, phase composition, corrosion and wear performance of plasma electrolytic oxidation (PEO) coatings on AM50 Mg was investigated. Different treatment durations were applied to fabricate coatings in an alkaline, phosphate-based electrolyte (1 g/L KOH + 20 g/L Na₃PO₄ + 5 g/L SiO₂), aiming to control the incorporated amount of SiO₂ particles in the layer. It was found that the uptake of particles was accompanied by the coating growth at the initial stage, while the particle content remained unchanged at the final stage, which is dissimilar to the evolution of the coating thickness. The incorporation mode of the particles and phase composition of the layer was not affected by the treatment duration under the voltage-control regime. The corrosion performance of the coating mainly depends on the barrier property of the inner layer, while wear resistance primarily relies on the coating thickness.

Keywords: magnesium; plasma electrolytic oxidation; SiO₂ particle; corrosion resistance; wear resistance

---

1. Introduction

Inferior corrosion and wear resistance are the main issues that restrict the wide range of applications of Mg and its alloys [1–4]. Plasma electrolytic oxidation (PEO) is one of the promising surface treatment processes derived from conventional anodizing to produce ceramic-like coatings on light alloys (Al, Mg and Ti) with enhanced anti-corrosion properties, wear resistance and biological compatibility [5–8]. In the case of Mg alloys, aqueous alkaline electrolytes are usually used during the PEO process and coatings are formed by the localized dielectric breakdown of the oxide film at high voltage [9–11]. The microstructure, phase composition and properties of the coating primarily depend on the electrolyte components and applied electrical parameters [12–14]. Particularly, energy input is the main driving force when producing a coating and plays an important role in the coating formation process. The current-control and voltage-control modes are generally applied to fabricate coatings under pulsed DC, AC or bipolar regimes [15–17].

Recently, solid particles (SiC, TiO₂, Al₂O₃, ZrO₂, etc.) have been introduced into PEO electrolyte to provide a wider range of phase compositions and new functionalities for PEO-coated Mg alloys [18–25].

It was found that the uptake and incorporation of the particles were significantly influenced by electrical parameters, such as the voltage/current density, frequency and duty cycle, leading to a modified coating microstructure, phase composition and properties [26–30]. Lin et al. [31] investigated the effects of voltage and oxidation duration on the corrosion properties of PEO coating using HA (hydroxyapatite)-containing electrolyte. It was proposed that higher voltage facilitates the incorporation of HA particles into the coating and enhances the corrosion resistance of the layer. However, a longer treatment duration induces more defects in the coating, which can act as a rapid route for the penetration of corrosive ions. Our previous work demonstrated that a lower frequency and higher duty ratio generates large-sized pores, leading to the uptake of more particles into the coating. Moreover, it is assumed that the pulse-on duration of each pulse is more important than the pulse-off duration for particle uptake during PEO processing [32]. To date, there have only been a limited number of reports on the effect of treatment duration on PEO coatings with addition of particles under a voltage-control regime. In the present work, the influence of SiO_2 particles on the microstructure, phase composition and properties of the coating was investigated to elucidate the optimized processing parameters to improve the coating properties.

2. Materials and Methods

Specimens of AM50 Mg alloy with dimensions of 15 mm × 15 mm × 4 mm were cut from gravity cast ingot material. The chemical composition of AM50 alloy as identified by an Arc Spark OES (Spark analyser M9, Spectro Analytical Instruments GmbH, Kleve, Germany) is 4.74 wt % Al, 0.383 wt % Mn, 0.065 wt % Zn, 0.063 wt %, Si, 0.002 wt % Fe, 0.002 wt % Cu and Mg balance. The specimens were ground using SiC abrasive papers up to 1200 grit, rinsed with ethanol and then air-dried prior to PEO treatment.

The PEO process was performed using a lab-produced pulsing unit in combination with a commercial power supply (PS 8000 2U, Elektro-Automatik, Viersen, Germany). The frequency was 250 Hz and the duty ratio were 10%. The specimen and a stainless steel tube were used as the anode and cathode, respectively. A total of 5 g/L of micro-sized SiO_2 particles were added to a phosphate-based electrolyte (20 g/L Na_3PO_4 and 1 g/L KOH). PEO coatings were produced under a constant voltage regime (450 V) for different treatment durations (1, 3, 5, 10, 20, 30 and 60 min). A stirrer and bubbling generator were used to facilitate the uniform distribution of the particles in the electrolyte. The temperature of the electrolyte was maintained at 20 ± 2 °C by a water cooling system.

A scanning electron microscope (SEM, TESCAN Vega3 SB, TESCAN, Brno, Czech Republic) combined with an energy dispersive spectrometer (EDS) system was used to examine the surface morphology, composition and microstructure of the PEO coatings. An acceleration voltage of 15 kV was applied for SEM and EDS investigations. The phase composition analysis was done with a Bruker X-ray diffractometer (XRD, Bruker, Billerica, MA, USA) using Cu Kα radiation. The dry sliding wear behavior of the PEO coatings was assessed using a Tribotec ball-on-disc oscillating tribometer (Tribotec, Brno, Czech Republic) with an AISI 52100 steel ball of 6 mm diameter as the static friction partner. The wear tests were performed under ambient conditions (25 ± 2 °C and 30% relative humidity) under a 5 N load with an oscillating amplitude of 10 mm at a sliding velocity of 5 mm s^{-1}, for a sliding distance of 12 m.

The corrosion behavior of the PEO coatings was assessed by electrochemical impedance spectroscopy (EIS) tests using an ACM Gill AC computer-controlled potentiostat (ACM Instruments, Cumbria, UK). A typical three-electrode cell with a saturated Ag/AgCl electrode as the reference electrode, a platinum mesh as the counter electrode and a coated specimen as the working electrode (0.5 cm^2 exposed area) was used. EIS studies were performed at open circuit potential (OCP) with an AC amplitude of 10 mV RMS (root mean square) sinusoidal perturbations over the frequency range from 30 kHz to 0.01 Hz. The measurements were taken at 0 (after 5 min immersion), 1, 3, 6, 12, 24, 48 and 72 h immersion time.

3. Results and Discussion

3.1. Microstructure

The effect of the treatment duration on the surface morphology of the PEO coatings is shown in Figure 1. It is evident that the number, size and distribution of the open pores is mainly related to the treatment duration. At the very beginning (1 min), few open pores appear in localized areas on the coating surface, indicating that the growth of the layer is inhomogeneous at the initial stage. A larger number of large-sized pores can be observed on the coating surface with a prolonged treatment duration (3 and 5 min). The specimen coated for 20 min revealed a much higher pore density with large-sized and uniformly distributed pores. It is also worth noting that the pore morphology on the coating surface remained unchanged after oxidation for 1 h. Moreover, apparent protrusions in the vicinity of large-sized pores indicate re-deposition of melted coating materials after the localized dielectric breakdown of the layer.

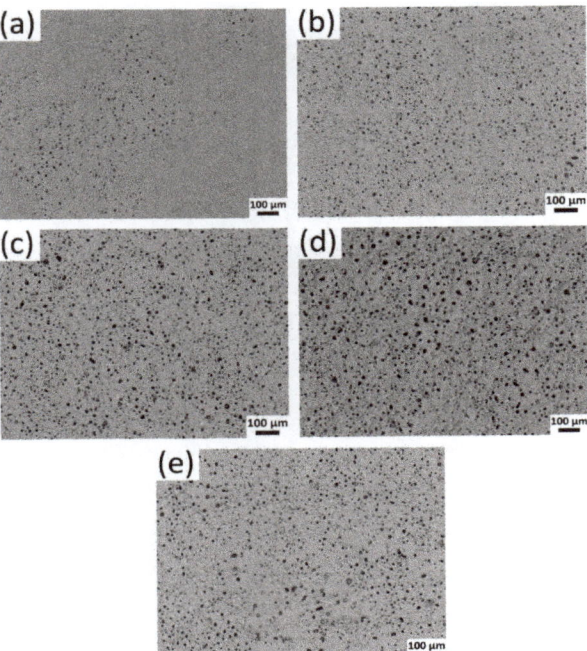

Figure 1. Backscattered electron images of surface morphology of plasma electrolytic oxidation (PEO) coatings produced after different treatment durations: (**a**) 1 min; (**b**) 3 min; (**c**) 5 min; (**d**) 20 min; (**e**) 60 min

Backscattered electron images of the cross-section of the coatings are shown in Figure 2. The coatings are composed of two layers, an outer porous layer and an inner barrier layer, between which a characteristic pore band for PEO coatings produced from phosphate-based electrolyte is clearly observed [33]. A longer treatment duration is likely to generate more defects in the inner layer, specifically for the coating treated for 60 min. The cross-section of the 1 min-treated PEO coating was very inhomogeneous and characterized by varied thicknesses in different regions, which is in good agreement with the pore morphology on the coating surface. The thickness of the coating increases with the treatment duration. Moreover, the coating thickness measured by SEM observation is well corroborated with the measurements from the eddy current probe (Figure 3). The growth of the PEO coatings exhibited significant variance during different stages. For example, the coatings grew at a rate of approximately 8 μm/min in the initial 5 min. However, the growth of the coatings slowed down to around 0.2 μm/min after reaching a certain thickness (40 μm). This can be ascribed to the low current density at the final stage of the voltage-control regime. A thick layer is harder to grow when the applied voltage remains unchanged [34,35]. As for the defects and open pores, they are slightly affected by the treatment duration and coating thickness.

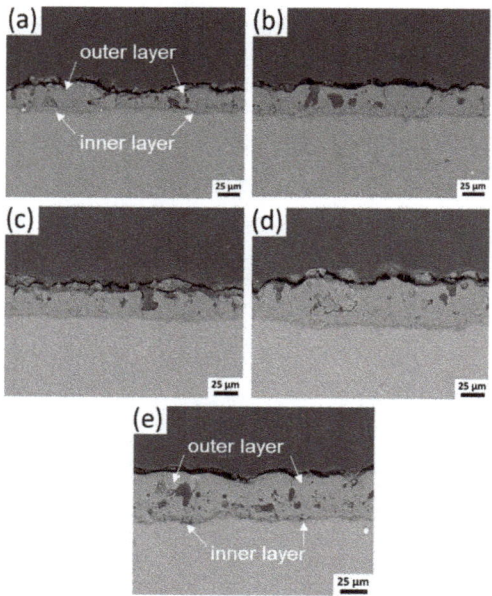

Figure 2. Backscattered electron images with a cross-sectional morphology of the plasma electrolytic oxidation (PEO) coatings produced after different treatment durations: (**a**) 1 min; (**b**) 3 min; (**c**) 5 min; (**d**) 20 min; (**e**) 60 min.

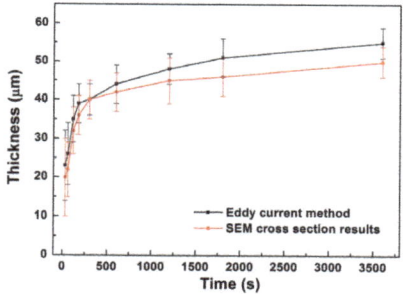

Figure 3. Thickness of the coatings treated for 1, 3, 5, 10, 20, 30 and 60 min.

3.2. Phase and Chemical Composition

The X-ray diffraction patterns of the coatings are shown in Figure 4. The appearance of Mg peaks is ascribed to the penetration of the X-ray through the entire coating. SiO_2 particles can be found in all the coatings except the one treated for 1 min, indicating that the micro-sized particles have been inertly incorporated into the layer and the coating treated for a short duration might be too thin to preserve the particles. Particularly, the coatings treated for 20 min and 1 h are composed of an amorphous phase in the 2θ range of 20–30°, possibly based on phosphorus-containing phases, which is consistent with our previous investigations [19]. EDS analysis was performed on the surface of all the coatings and the Si content is depicted in Figure 5, showing particle uptake during the coating growth process. It was found that the amount of the incorporated particles increased rapidly within the first 5 min, then stayed at the same level (6.3 at.%) in the later stage. It is worth noting that the evolution of the particle content on the coating surface is somehow dissimilar to that of the coating thickness at the final stage. The layer grew slowly outwards after a certain treatment duration, while the particle content remained unchanged in the meantime. This can be ascribed to the fact that the uptake of particles is primarily related to the characteristics of the discharges and pore morphology on the coating surface, which remain almost unchanged at the final stage under a constant voltage mode.

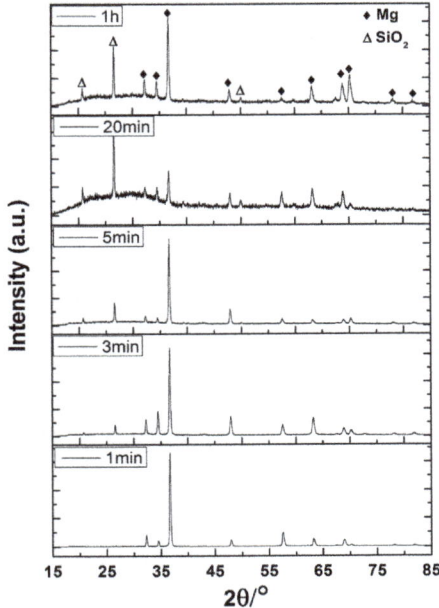

Figure 4. XRD patterns of the coatings treated for 1, 3, 5, 20 and 60 min.

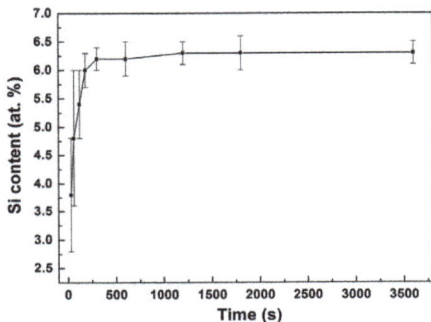

Figure 5. Si content of the coatings treated for 1, 3, 5, 10, 20, 30 and 60 min.

3.3. Corrosion Behavior

The degradation behavior of the coatings in 0.5 wt % NaCl solution was examined by EIS measurements. The Bode plots of the EIS spectra for PEO coatings obtained after different treatment durations are presented in Figure 6. Two well-defined time constants can be distinguished in all Bode plots at low and high frequencies, except for the coating treated for 1 min. It should be noted that this measurement was done after immersion for 5 min when the open circuit potential (OCP) of the coating was relatively stable. For the specimen treated for 1 min, the coating was too thin and inhomogeneous to prevent the substrate from corrosion. However, the aggressive ions could not fill the open pores and reach the substrate in the case of thick coatings. Therefore, the time constant at high frequency (10^4 Hz) can be assigned to the outer layer, while the time constant at lower frequencies (10 Hz) can be assigned to the compact inner layer. Signs of an additional time constant in the low frequency range appear after 1-h immersion. This is ascribed to the initiation of the corrosion process at the metal/electrolyte interface. The time constant at high frequencies disappeared with a prolonged immersion time, indicating that the outer layer is fully penetrated by the electrolyte.

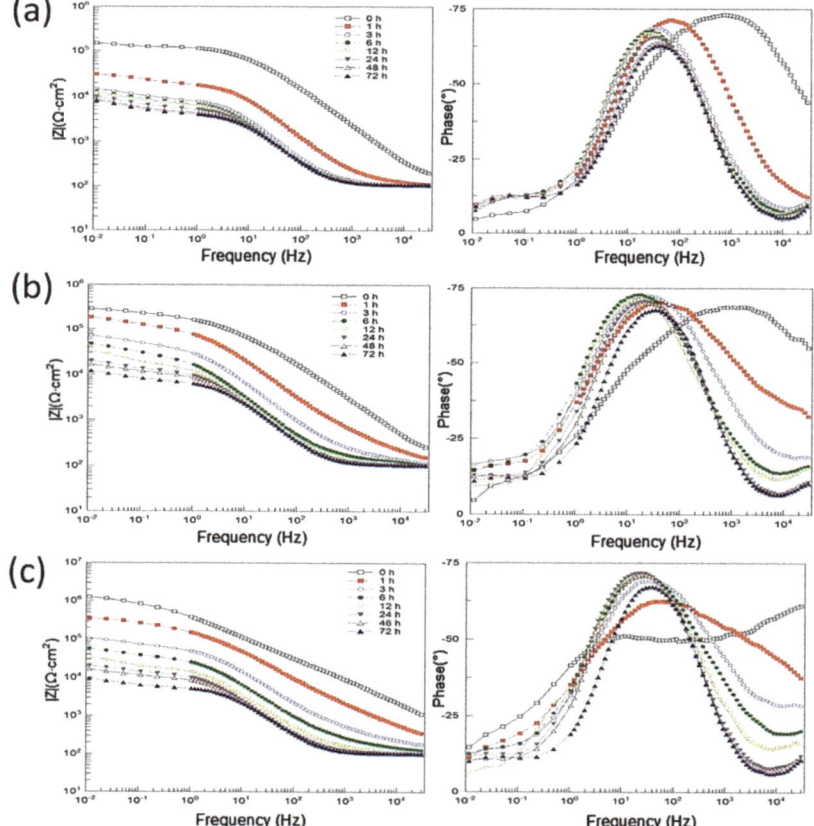

Figure 6. *Cont.*

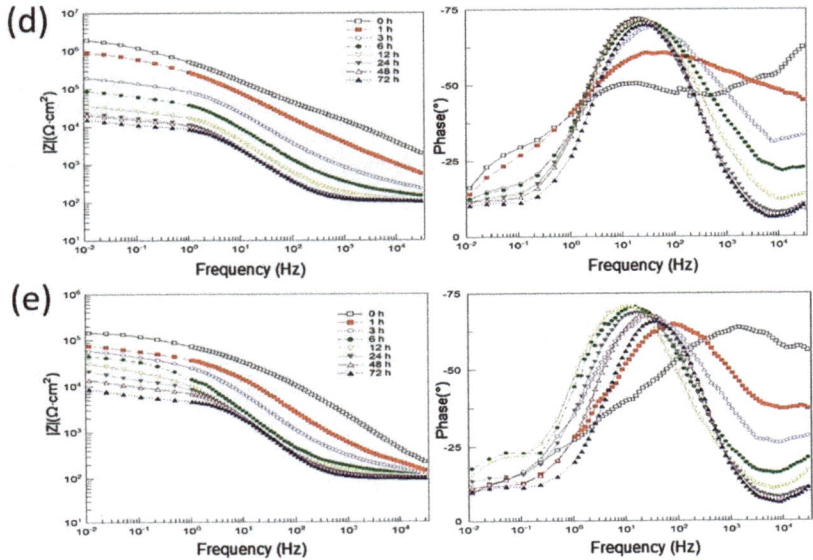

Figure 6. Electrochemical impedance behavior (bode plots) of the PEO coatings produced after different treatment durations: (**a**) 1 min; (**b**) 3 min; (**c**) 5 min; (**d**) 20 min; (**e**) 60 min.

The impedance at the lowest frequency (0.01 Hz) is used to roughly estimate the total corrosion resistance of different systems (Figure 7) to manifest the influence of treatment duration on the corrosion behavior of the PEO coatings. As expected, the coating treated for 1 min exhibited the lowest corrosion resistance and fastest degradation of all the coatings. The coating treated for 20 min showed the highest corrosion resistance, while the thickest coating (oxidized for 60 min) demonstrated relatively inferior property. The corrosion resistance of the coating treated for 3, 5 and 20 min was quite different, while the Si content on the coating surface was nearly the same. Therefore, the presence of the particles had a slight influence on the corrosion resistance of the PEO coating. The corrosion performance and degradation behavior of the PEO coating are mainly related to the coating thickness and barrier property of the inner layer, while the latter is more critical for the corrosion performance.

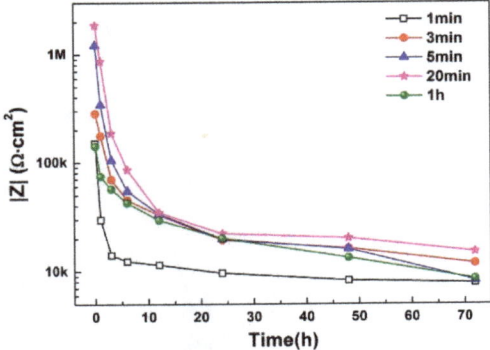

Figure 7. The change of $|Z|_f$ = 0.01 Hz of the PEO coatings produced after different treatment durations (1 min, 3 min, 5 min, 20 min, and 60 min) as a function of immersion time.

3.4. Tribological Performance

The evolution of the friction coefficient determined for the respective coatings is shown in Figure 8. In the first 3 m of sliding, the friction coefficient rose rapidly to 0.76 for the coating treated for 3 min, followed by large fluctuations in the range of 0.42–0.75, indicating the failure of the coating and exposure of the Mg substrate. A similar behavior can be observed for the coating treated for 5 min. In the case of the coatings treated for a longer duration, the friction coefficient was relatively low, without severe fluctuation during the entire wear test.

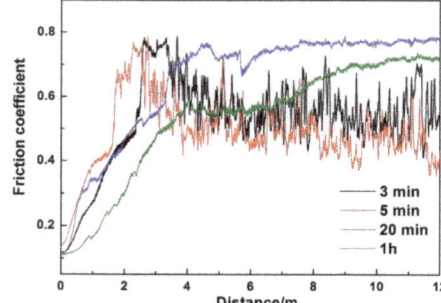

Figure 8. The evolution of the friction coefficient of the PEO coatings anodized for different durations in the dry sliding test.

The wear tracks of the coatings after the dry sliding wear test are shown in Figure 9. The wear tracks of the coating treated for 3 and 5 min (Figure 9a,b) are broad and deep, where many grooves, parallel scratches and oxidized regions are visible. This indicates that the thin coatings are completely removed, corresponding to the recorded friction coefficient profile (Figure 8). On the contrary, the wear track of the coatings treated for a longer treatment duration (20 min and 1 h) are still intact and compact, which is consistent with the EDS analysis of the wear tracks (Table 1). The enhanced tribological performance can be ascribed to the thick layer with inertly incorporated SiO_2 particles on the coating surface.

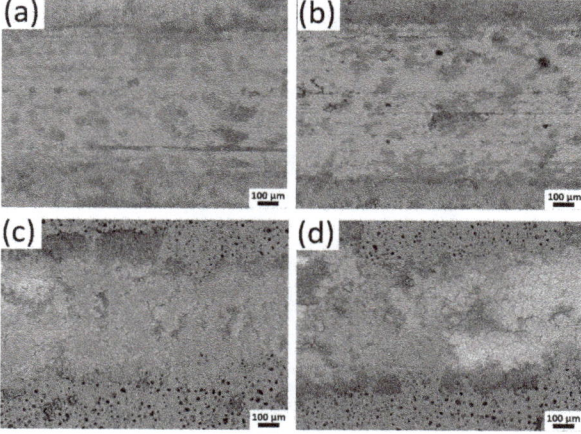

Figure 9. Surface morphology of the corresponding counterparts (steel ball) after wear test: PEO coatings treated for (**a**) 3 min, (**b**) 5 min, (**c**) 20 min, and (**d**) 60 min.

Table 1. Elemental composition (at.%) of the wear tracks in Figure 9.

Coating	O	Na	Mg	Al	P	Si
PEO (3 min)	8	–	90.5	0.5	1	–
PEO (5 min)	10	–	88.5	0.5	1	–
PEO (20 min)	60	4.5	19	1	9	6.5
PEO (60 min)	59	5	20	1	9	6

4. Conclusions

- The amount of particles increased significantly in the initial stage of coating growth, while it remained unchanged in the final stage under a constant voltage regime, which is dissimilar to the evolution of the coating thickness.
- The corrosion performance and wear performance of the coating is primarily related to the coating thickness and barrier property of the inner layer. Of these, the corrosion resistance of the coating mainly depended on the barrier property of the inner layer, while the wear resistance primarily relied on the coating thickness.
- The addition of SiO_2 particles can greatly enhance the wear resistance of the PEO coating.

Author Contributions: Formal Analysis, X.L., Y.C. and Y.L.; Investigation, X.L.; Methodology, Y.L.; Resources, F.W. and K.U.K.; Supervision, C.B., K.U.K. and M.Z.; Validation, T.Z.; Writing—Original Draft Preparation, X.L. and Y.C.; Writing—Review & Editing, C.B., T.Z., F.W. and M.Z.

Funding: This research was funded by the National Natural Science Foundation of China (U1737102, 51531007 and 51771050), Young Elite Scientists Sponsorship Program by CAST (2017QNRC001), Fundamental Research Funds for the Central Universities (N170203006 and N170205002) and National Program for Young Top-notch Professionals.

Acknowledgments: The technical support of Volker Heitmann and Ulrich Burmester is gratefully acknowledged.

Conflicts of Interest: The authors declare no conflict of interest.

References

1. Yang, J.; Lu, X.; Blawert, C.; Di, S.; Zheludkevich, M.L. Microstructure and corrosion behavior of Ca/P coatings prepared on magnesium by plasma electrolytic oxidation. *Surf. Coat. Technol.* **2017**, *319*, 359–369. [CrossRef]
2. Lu, X.; Chen, Y.; Zhang, C.; Zhang, T.; Yu, B.; Xu, H.; Wang, F. Formation mechanism and corrosion performance of phosphate conversion coatings on AZ91 and Mg–Gd–Y–Zr alloy. *J. Electrochem. Soc.* **2018**, *165*, C601–C607. [CrossRef]
3. Wu, L.; Yang, D.; Zhang, G.; Zhang, Z.; Zhang, S.; Tang, A.; Pan, F. Fabrication and characterization of Mg-M layered double hydroxide films on anodized magnesium alloy AZ31. *Appl. Surf. Sci.* **2018**, *431*, 177–186. [CrossRef]
4. Yang, L.; Liu, G.; Ma, L.; Zhang, E.; Zhou, X.; Thompson, G. Effect of iron content on the corrosion of pure magnesium: Critical factor for iron tolerance limit. *Corros. Sci.* **2018**, *139*, 421–429. [CrossRef]
5. Aliofkhazraei, M.; Gharabagh, R.S.; Teimouri, M.; Ahmadzadeh, M.; Darband, G.B.; Hasannejad, H. Ceria embedded nanocomposite coating fabricated by plasma electrolytic oxidation on titanium. *J. Alloys Compd.* **2016**, *685*, 376–383. [CrossRef]
6. Lu, X.; Mohedano, M.; Blawert, C.; Matykina, E.; Arrabal, R.; Kainer, K.U.; Zheludkevich, M.L. Plasma electrolytic oxidation coatings with particle additions—A review. *Surf. Coat. Technol.* **2016**, *307*, 1165–1182. [CrossRef]
7. Mohedano, M.; Serdechnova, M.; Starykevich, M.; Karpushenkov, S.; Bouali, A.C.; Ferreira, M.G.S.; Zheludkevich, M.L. Active protective PEO coatings on AA2024: Role of voltage on in-situ LDH growth. *Mater. Des.* **2017**, *120*, 36–46. [CrossRef]
8. Matykina, E.; Garcia, I.; Arrabal, R.; Mohedano, M.; Mingo, B.; Sancho, J.; Merino, M.C.; Pardo, A. Role of PEO coatings in long-term biodegradation of a Mg alloy. *Appl. Surf. Sci.* **2016**, *389*, 810–823. [CrossRef]

9. Chen, Y.; Lu, X.; Blawert, C.; Zheludkevich, M.L.; Zhang, T.; Wang, F. Formation of self-lubricating PEO coating via in-situ incorporation of PTFE particles. *Surf. Coat. Technol.* **2018**, *337*, 379–388. [CrossRef]
10. Lu, X.; Blawert, C.; Tolnai, D.; Subroto, T.; Kainer, K.U.; Zhang, T.; Wang, F.; Zheludkevich, M.L. 3D reconstruction of plasma electrolytic oxidation coatings on Mg alloy via synchrotron radiation tomography. *Corros. Sci.* **2018**, *139*, 395–402. [CrossRef]
11. Mohedano, M.; Lu, X.; Matykina, E.; Blawert, C.; Arrabal, R.; Zheludkevich, M.L. Plasma electrolytic oxidation (PEO) of metals and alloys. In *Encyclopedia of Interfacial Chemistry*, 1st ed.; Wandelt, K., Ed.; Elsevier: Oxford, UK, 2018; pp. 423–438.
12. Song, Y.; Dong, K.; Shan, D.; Han, E.-H. Investigation of a novel self-sealing pore micro-arc oxidation film on AM60 magnesium alloy. *J. Magnes. Alloys* **2013**, *1*, 82–87. [CrossRef]
13. Dong, K.; Song, Y.; Shan, D.; Han, E.-H. Corrosion behavior of a self-sealing pore micro-arc oxidation film on AM60 magnesium alloy. *Corros. Sci.* **2015**, *100*, 275–283. [CrossRef]
14. Stojadinović, S.; Vasilić, R.; Radić-Perić, J.; Perić, M. Characterization of plasma electrolytic oxidation of magnesium alloy AZ31 in alkaline solution containing fluoride. *Surf. Coat. Technol.* **2015**, *273*, 1–11. [CrossRef]
15. Yu, L.; Cao, J.; Cheng, Y. An improvement of the wear and corrosion resistances of AZ31 magnesium alloy by plasma electrolytic oxidation in a silicate–hexametaphosphate electrolyte with the suspension of SiC nanoparticles. *Surf. Coat. Technol.* **2015**, *276*, 266–278. [CrossRef]
16. Barati Darband, G.; Aliofkhazraei, M.; Hamghalam, P.; Valizade, N. Plasma electrolytic oxidation of magnesium and its alloys: Mechanism, properties and applications. *J. Magnes. Alloys* **2017**, *5*, 74–132. [CrossRef]
17. Cui, L.-Y.; Gao, S.-D.; Li, P.-P.; Zeng, R.-C.; Zhang, F.; Li, S.-Q.; Han, E.-H. Corrosion resistance of a self-healing micro-arc oxidation/polymethyltrimethoxysilane composite coating on magnesium alloy AZ31. *Corros. Sci.* **2017**, *118*, 84–95. [CrossRef]
18. Lu, X.; Blawert, C.; Zheludkevich, M.L.; Kainer, K.U. Insights into plasma electrolytic oxidation treatment with particle addition. *Corros. Sci.* **2015**, *101*, 201–207. [CrossRef]
19. Lu, X.; Blawert, C.; Huang, Y.; Ovri, H.; Zheludkevich, M.L.; Kainer, K.U. Plasma electrolytic oxidation coatings on Mg alloy with addition of SiO_2 particles. *Electrochim. Acta* **2016**, *187*, 20–33. [CrossRef]
20. Zoubi, W.A.; Kamil, M.P.; Ko, Y.G. Synergistic influence of inorganic oxides (ZrO_2 and SiO_2) with N_2H_4 to protect composite coatings obtained via plasma electrolyte oxidation on Mg alloy. *Phys. Chem. Chem. Phys.* **2017**, *19*, 2372–2382. [CrossRef] [PubMed]
21. Stojadinović, S.; Tadić, N.; Radić, N.; Grbić, B.; Vasilić, R. MgO/ZnO coatings formed on magnesium alloy AZ31 by plasma electrolytic oxidation: Structural, photoluminescence and photocatalytic investigation. *Surf. Coat. Technol.* **2017**, *310*, 98–105. [CrossRef]
22. Mohedano, M.; Arrabal, R.; Mingo, B.; Pardo, A.; Matykina, E. Role of particle type and concentration on characteristics of PEO coatings on AM50 magnesium alloy. *Surf. Coat. Technol.* **2018**, *334*, 328–335. [CrossRef]
23. NasiriVatan, H.; Ebrahimi-Kahrizsangi, R.; Asgarani, M.K. Tribological performance of PEO-WC nanocomposite coating on Mg alloys deposited by plasma electrolytic oxidation. *Tribol. Int.* **2016**, *98*, 253–260. [CrossRef]
24. Pezzato, L.; Angelini, V.; Brunelli, K.; Martini, C.; DabalÀ, M. Tribological and corrosion behavior of PEO coatings with graphite nanoparticles on AZ91 and AZ80 magnesium alloys. *Trans. Nonferrous Met. Soc. China* **2018**, *28*, 259–272. [CrossRef]
25. Tonelli, L.; Pezzato, L.; Dolcet, P.; Dabalà, M.; Martini, C. Effects of graphite nano-particle additions on dry sliding behaviour of plasma-electrolytic-oxidation-treated EV31A magnesium alloy against steel in air. *Wear* **2018**, *404*, 122–132. [CrossRef]
26. Lu, X.; Blawert, C.; Kainer, K.U.; Zheludkevich, M.L. Investigation of the formation mechanisms of plasma electrolytic oxidation coatings on Mg alloy AM50 using particles. *Electrochim. Acta* **2016**, *196*, 680–691. [CrossRef]
27. Shokouhfar, M.; Allahkaram, S.R. Formation mechanism and surface characterization of ceramic composite coatings on pure titanium prepared by micro-arc oxidation in electrolytes containing nanoparticles. *Surf. Coat. Technol.* **2016**, *291*, 396–405. [CrossRef]

28. Yeung, W.K.; Sukhorukova, I.V.; Shtansky, D.V.; Levashov, E.A.; Zhitnyak, I.Y.; Gloushankova, N.A.; Kiryukhantsev-Korneev, P.V.; Petrzhik, M.I.; Matthews, A.; Yerokhin, A. Characteristics and in vitro response of thin hydroxyapatite-titania films produced by plasma electrolytic oxidation of Ti alloys in electrolytes with particle additions. *RSC Adv.* **2016**, *6*, 12688–12698. [CrossRef] [PubMed]
29. Liu, C.-Y.; Tsai, D.-S.; Wang, J.-M.; Tsai, J.T.J.; Chou, C.-C. Particle size influences on the coating microstructure through green chromia inclusion in plasma electrolytic oxidation. *ACS Appl. Mater. Interfaces* **2017**, *9*, 21864–21871. [CrossRef] [PubMed]
30. Lou, B.-S.; Lin, Y.-Y.; Tseng, C.-M.; Lu, Y.-C.; Duh, J.-G.; Lee, J.-W. Plasma electrolytic oxidation coatings on AZ31 magnesium alloys with Si_3N_4 nanoparticle additives. *Surf. Coat. Technol.* **2017**, *332*, 358–367. [CrossRef]
31. Lin, X.; Wang, X.; Tan, L.; Wan, P.; Yu, X.; Li, Q.; Yang, K. Effect of preparation parameters on the properties of hydroxyapatite containing micro-arc oxidation coating on biodegradable ZK60 magnesium alloy. *Ceram. Int.* **2014**, *40*, 10043–10051. [CrossRef]
32. Lu, X.; Blawert, C.; Mohedano, M.; Scharnagl, N.; Zheludkevich, M.L.; Kainer, K.U. Influence of electrical parameters on particle uptake during plasma electrolytic oxidation processing of AM50 Mg alloy. *Surf. Coat. Technol.* **2016**, *289*, 179–185. [CrossRef]
33. Lu, X.; Sah, S.P.; Scharnagl, N.; Störmer, M.; Starykevich, M.; Mohedano, M.; Blawert, C.; Zheludkevich, M.L.; Kainer, K.U. Degradation behavior of PEO coating on AM50 magnesium alloy produced from electrolytes with clay particle addition. *Surf. Coat. Technol.* **2015**, *269*, 155–169. [CrossRef]
34. Hussein, R.O.; Nie, X.; Northwood, D.O. An investigation of ceramic coating growth mechanisms in plasma electrolytic oxidation (PEO) processing. *Electrochim. Acta* **2013**, *112*, 111–119. [CrossRef]
35. Cheng, Y.-L.; Xue, Z.-G.; Wang, Q.; Wu, X.-Q.; Matykina, E.; Skeldon, P.; Thompson, G.E. New findings on properties of plasma electrolytic oxidation coatings from study of an Al–Cu–Li alloy. *Electrochim. Acta* **2013**, *107*, 358–378. [CrossRef]

© 2018 by the authors. Licensee MDPI, Basel, Switzerland. This article is an open access article distributed under the terms and conditions of the Creative Commons Attribution (CC BY) license (http://creativecommons.org/licenses/by/4.0/).

Article

Porous CaP Coatings Formed by Combination of Plasma Electrolytic Oxidation and RF-Magnetron Sputtering

Anna Kozelskaya [1,*], Gleb Dubinenko [1], Alexandr Vorobyev [1], Alexander Fedotkin [1], Natalia Korotchenko [2], Alexander Gigilev [2], Evgeniy Shesterikov [3,4], Yuriy Zhukov [5] and Sergei Tverdokhlebov [1,*]

1. School of Nuclear Science & Engineering, Tomsk Polytechnic University, 634050 Tomsk, Russia; dubinenko.gleb@gmail.com (G.D.); alexandr.vorobyev13@gmail.com (A.V.); fedotkin_sasha@mail.ru (A.F.)
2. Department of Inorganic Chemistry, Tomsk State University, 634050 Tomsk, Russia; korotch@mail.ru (N.K.); gigilev@mail.tsu.ru (A.G.)
3. Laboratory of Radiophotonics, V.E. Zuev Institute of Atmospheric Optics SB RAS, 634055 Tomsk, Russia; shesterikov_e@mail.ru
4. Nanotechnology Center, Tomsk State University of Control Systems and Radioelectronics, 634050 Tomsk, Russia
5. Department of Nuclear-Physics Research Methods, Saint-Petersburg State University, 199034 Saint-Petersburg, Russia; YURI.ZHUKOV@prevac.ru
* Correspondence: kozelskayaai@tpu.ru (A.K.); tverd@tpu.ru (S.T.)

Received: 28 October 2020; Accepted: 17 November 2020; Published: 19 November 2020

Abstract: The porous CaP subcoating was formed on the Ti6Al4V titanium alloy substrate by plasma electrolytic oxidation (PEO). Then, upper coatings were formed by radio frequency magnetron sputtering (RFMS) over the PEO subcoating by the sputtering of various CaP powder targets: β-tricalcium phosphate (β-TCP), hydroxyapatite (HA), Mg-substituted β-tricalcium phosphate (Mg-β-TCP) and Mg-substituted hydroxyapatite (Mg-HA), Sr-substituted β-tricalcium phosphate (Sr-β-TCP) and Sr-substituted hydroxyapatite (Sr-HA). The coating surface morphology was studied by scanning electron and atomic force microscopy. The chemical composition was determined by X-ray photoelectron spectroscopy. The phase composition of the coatings was studied by X-ray diffraction analysis. The Young's modulus of the coatings was studied by nanoindentation test. RF-magnetron sputtering treatment of PEO subcoating resulted in multileveled roughness, increased Ca/P ratio and Young's modulus and enrichment with Sr and Mg. Sputtering of the upper layer also helped to adjust the coating crystallinity.

Keywords: plasma electrolytic oxidation (PEO); radio frequency magnetron sputtering (RFMS); calcium-phosphate (CaP) coating

1. Introduction

Implant integration on the bone site, together with the bone fracture consolidation time and effectiveness, are among the most significant conditions for effective bone healing. Bioactive calcium phosphate (CaP) coatings are assigned to improve metallic implant osseointegration and reduce bone healing time. Micro-arc oxidation (MAO) or plasma electrolytic oxidation (PEO) and radio frequency magnetron sputtering deposition (RFMS) are the most developed CaP coating deposition methods.

Compared to other surface modification methods, PEO is economically efficient and best for the deposition of bioactive coatings with open porosity [1,2]. PEO coatings are excellent for their high wear and corrosive resistance [3–6], and low residual stress on the surface due to porous morphology [7]. Porous morphology of the coating enhances implant osseointegration and promotes bone tissue regeneration.

Moreover, the raw surface provides a higher available surface area for cell seeding and bone morphogenetic protein (BMP) adsorption [8]. Another PEO feature is the feasibility of modifying the surface of complex shape and 3D-printed implantable devices. In turn, RF-sputtered coatings are dense and known for their higher calcium-to-phosphorus (Ca/P) ratio. RFMS allows for coating deposition on the various types of materials, including ceramics and polymers. This method is variable not only due to substrate material options, but also due to the material of the sputtered target.

RFMS and PEO are also known for their disadvantages. PEO coatings can only be deposited on the surface of the valve metals [9–12]. RFMS requires expensive vacuum equipment and restricted by a low rate of deposition, which determines the low efficiency of this method [13,14]. Moreover, magnetron sputtering cannot ensure coating deposition on the inner surfaces of complex-shaped substrates. A combination of PEO and RFMS could consolidate their advantages and compensate for the drawbacks.

A combination of PEO and RFMS was used previously [15,16]. Park et al., revealed the effect of enhanced corrosion potential and passive current density after RFMS of Mn on the Ti–29Nb–xHf substrate [15]. The molar ratio of (Ca + Mn)/P was growing with the increase in Mn sputtering time. Hwang et al. combined PEO and RFMS to fabricate CaP/Zn coatings on the Ti–6Al–4V [16]. Coatings RFMS with Zn showed a higher cell proliferation rate. Extended Zn RFMS leads to an increased (Ca + Zn)/P ratio, growth of the amorphous phase and increase in Zn corrosion rate.

In the presented study, we sputtered thin RFMS CaP coatings of various compositions to enhance the osteostimulating properties and increase the Ca/P ratio of the PEO CaP subcoating. An increase in Ca/P ratio, the formation of multileveled roughness and enrichment with Mg and Sr are engaged to potentially regulate osteoblasts' and osteoclasts' functionality and enhance cell attachment and proliferation [17–20].

2. Materials and Methods

2.1. Coatings Formation

Ti–6Al–4V 10 mm in diameter and 1-mm-thick disks were used as substrates for coatings formation. Preparation of the surface of the samples before coating included cleaning in an ultrasonic bath in distilled water and chemical etching in an aqueous solution of nitric and hydrofluoric acids taken in volume ratios of $HNO_3:HF:H_2O$ = 1:2.5:2.5, at the temperature of 15–20 °C for 10–15 s, followed by neutralization in a 1% aqueous solution of sodium hydroxide and repeated washing with distilled water.

PEO subcoating was formed on the "Micro-arc oxidation complex" (PEOC) designed in the Laboratory for Plasma Hybrid Systems, The Weinberg Research Center, School of Nuclear Science & Engineering of Tomsk Polytechnic University (Tomsk, Russia). A supersaturated solution of CaO in 10% H_3PO_4 with 10 g/L dispersed hydroxyapatite (particle size up to 70 μm) was used as an electrolyte. The density and pH of electrolyte were 1080–1090 kg/m^3 and 2.03, correspondingly. PEO subcoating was formed at the following operating mode: voltage—320 V, voltage rise rate—3 V/s, pulse repetition rate—200 Hz, pulse duration 100 μs, coating formation time—15 min. The temperature of electrolyte was keeping about 15 °C by cooling the bath by water flow. Coated samples were washed in distilled water and dried at 120 °C on air for 30 min.

Various RF magnetron sputtered upper coatings were formed by the sputtering of six different powder targets: pure β-tricalcium phosphate (β-TCP), pure hydroxyapatite (HA), Mg-substituted β-tricalcium phosphate and Mg-substituted hydroxyapatite (Mg-β-TCP, Mg-HA, Mg substitutions concentration was 1.53 wt.% ± 0.01 wt.%), Sr-substituted β-tricalcium phosphate and Sr-substituted hydroxyapatite (Sr-β-TCP, Sr-HA, Sr substitutions concentration was 3.39 wt.% ± 0.09 wt.%). β-TCP, Mg-β-TCP, Sr-β-TCP powders were synthesized in Riga Technical University (Riga, Latvia) [13]. HA-based powders were synthesized by microwave assisted method in Tomsk State University (Tomsk, Russia) [14]. Liquid phase synthesis of MgHA and SrHA powders was carried out using a stoichiometric ratio Ca/P = 1.67 ((Ca + Me)/P = 1.67) and the following reactions:

$$(10-x)Ca(NO_3)_2 + 6(NH_4)_2HPO_4 + xMg(NO_3)_2 + 8NH_4OH = Ca_{(10-x)}Mg_x(PO_4)_6(OH)_2 + 6H_2O + 20NH_4NO_3 \quad (1)$$

$$(10-x)Ca(NO_3)_2 + 6(NH_4)_2HPO_4 + xSr(NO_3)_2 + 8NH_4OH = Ca_{(10-x)}Sr_x(PO_4)_6(OH)_2 + 6H_2O + 20NH_4NO_3 \quad (2)$$

where x = 0.1; 0.3; 0.5 (mol), Me—metal.

An aqueous solution of calcium nitrate was mixed with a solution of ammonium hydrophosphate in concentration 0.5 and 0.3 M, respectively. The weighed sample of magnesium or strontium nitrate was added to the calcium nitrate solution. A reactant solution pH value of 10–11 was reached with an aqueous solution of ammonia (25%, ρ = 0.9 g/mL). The mixture underwent microwave exposure with 110 W for 40 min and then was carried at room temperature for 48 h. The precipitate was filtered, rinsed with a diluted solution of ethanol and dried until constant weight (~15 h) at 95 °C. Dried powders were annealed for 4 h at 900 °C.

Targets were formed by spreading powders in the hexagonal crucible with an area of 230 cm^2 and a depth of 0.6 cm. The volume was constant for all powders and equal to approximately 138 cm^3. The preliminary pressure in the chamber was 10^{-3} Pa, and the working pressure (Ar) was 0.5 Pa. The distance between the sputtering target and the substrate was 40 mm, power density—4.8 W/cm^2, sputtering time—7 h for HA-based powders and 21 h for β-TCP-based powders.

2.2. Coating Thickness

PEO sub-coating thickness was studied by the eddy current testing with the use of Konstanta 5 (KONSTANTA, Saint-Petersburg, Russia) equipment. RFMS upper coating was studied on the silicon witness samples with the use of Ellips 1891 SAG (NPK "Nanotechnology Center", Novosibirsk, Russia) spectral ellipsometer.

2.3. Scanning Electron Microscopy (SEM)

Coating morphology was studied on the scanning electron microscope JCM-6000 (JEOL, Tokyo, Japan) using a backscattered electron detector in a low vacuum mode. Micrographs were obtained at 15 kV accelerating voltage and 3.5 µA beam current. For better micrograph quality, all samples were preliminarily sputter-coated with gold using the SC7640 magnetron sputtering system (Quorum Technologies Ltd., Newhaven, UK).

2.4. Atomic Force Microscopy (AFM)

The surface morphologies of the coatings were examined using atomic force microscope (Solver-HV, NT-MDT, Moscow, Russia) operating in contact mode.

2.5. X-ray Photoelectron Spectroscopy (XPS)

The chemical composition of the coatings was studied using an Escalab 250Xi instrument (Thermo Fisher Scientific Inc., Saint-Petersburg, Russia) equipped with monochromatic AlKα radiation (photon energy 1486.6 eV). Total energy resolution was about 0.55 eV. Spectra were recorded in the constant pass energy mode at 100 eV for survey spectrum and 50 eV for element core level spectrum, using an XPS spot size of 650 µm. Investigations were carried out at room temperature in ultrahigh vacuum of 1×10^{-9} mbar (in case of the use of electron-ion compensation system, Ar partial pressure was 1×10^{-7} mbar). The library of the reference XPS spectra including the atomic registration sensitivity factors was provided by the instrument manufacturer within the Avantage Data System (Thermo Fisher Scientific Inc.). Peaks were deconvoluted using the Avantage software (version 5.977) which was set to a Shirley background subtraction followed by peak fitting to Voigt functions having 80% Gaussian and 20% Lorentzian character.

2.6. X-ray Diffraction (XRD) Analysis

Coating phase composition was studied with the use of Shimadzu XRD 6000 (Shimadzu, Tokyo, Japan) X-ray diffractometer using CuKα radiation. Diffraction patterns were obtained in Bragg–Brentano geometry with 40 kV accelerating voltage, 30 mA beam current, in the range of 10°–80° scanning angles with 1°/min scanning speed, 0.02° scanning step and 1 s signal acquisition time.

2.7. Nanoindentation

Composite coating Young's modulus was measured with the use of Nanoindenter G200 (Agilent's Electronic Measurement, Santa Clara, CA, USA). Indentation was performed with triangular Berkovich pyramid at 50 mN at least five times on each sample. Young's modulus was defined according to the ISO14577 [21].

3. Results

Two consecutive layers formed composition CaP coatings: thick PEO subcoating and thin RFMS upper coating of six different compositions. The thickness of each layer and total composite coating thickness are presented in Table 1. We observed no significant difference in the thickness of PEO subcoating on all samples, which was approximately 35 µm. However, the thickness of RFMS upper coating depends on the target composition and ion substitutions. β-TCP-based targets have a lower rate of sputtering comparing with HA-based targets. Another key factor affects the rate of sputtering is Mg and Sr ion substitutions. Sr substitution increases the rate of β-TCP RFMS, while Mg has no significant effect on the sputtering rate. We discussed the effect of Sr and Mg substitutions on the β-TCP RFMS rate in more detail in our previous study [22]. For HA-based upper coatings, we observed no significant effect of Sr and Mg substitution on the rate of sputtering. We revealed significant thickness differences with PEO only for the PEO + HA composite coating.

Table 1. Coating thickness.

Group	PEO Subcoating Thickness, µm	RFMS Upper Coating Thickness, µm	Total Composite Coating Thickness, µm
PEO	35 ± 1.7	–	35.0 ± 1.7
PEO + HA	36 ± 1.0	2.4 ± 0.1	38.4 ± 1.1 *
PEO + Sr-HA	36 ± 1.0	1.3 ± 0.1 †	37.3 ± 1.1
PEO + Mg-HA	34 ± 0.5	1.1 ± 0.1 †	35.1 ± 0.6
PEO + β-TCP	35 ± 1.5	0.7 ± 0.1	35.7 ± 1.6
PEO + Sr-β-TCP	36 ± 1.5	1.0 ± 0.1 †	37.0 ± 1.6
PEO + Mg-β-TCP	35 ± 1.5	0.6 ± 0.1	35.6 ± 1.6

* significant differences with PEO ($p < 0.05$); † significant differences between groups of the β-TCP and X-β-TCP coatings and HA and X-HA coatings, where X is Mg or Sr ($p < 0.05$).

3.1. Composite Coating Morphology

Close-standing highly porous groups of spheroidal structures (spherulites) form the surface of PEO subcoating (Figure 1a). The average spherulites lateral dimension varies from 5 to 20 µm and open spherulites walls thickness from 1 to 1.5 µm. We suppose that high-temperature microplasma discharges during PEO initiate erosion and partial destruction of spherulites, which results in the formation of a sponge-like structure. On the macroscale, this sponge structure is stable after RFMS of the upper coating. Thin upper coating replicates the subcoating macrorelief. However, at high magnification, we observe crystallite-like structures on the spherulites' surfaces (Figure 1b–g). We observed a similar surface morphology for composition coatings with β-TCP-based RFMS upper coatings (Figure 1e–g). It should be noted that distinct dendritic crystallites cover the surface of spherulites with HA upper coating (Figure 1b). We associate this crystallization effect with an intense surface heating due to bombarding substrate by high-energy plasma particles during RFMS of the HA target.

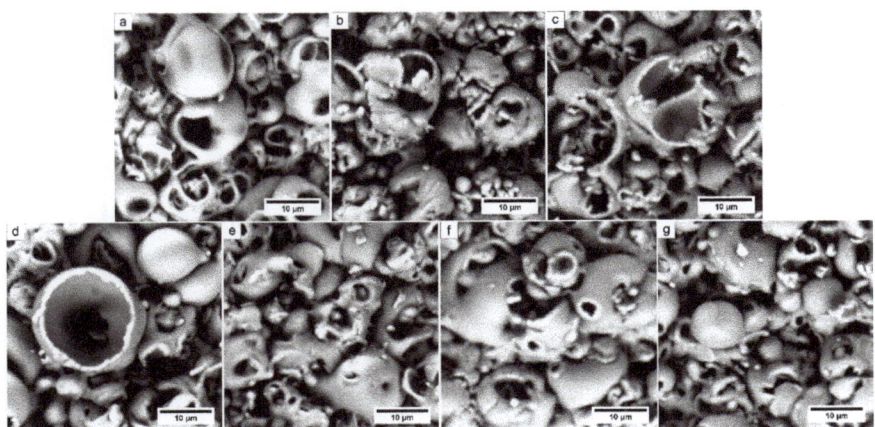

Figure 1. SEM micrographs of PEO CaP subcoating (**a**) and PEO + RFMS HA (**b**), Sr-HA (**c**), Mg-HA (**d**), β-TCP (**e**), Sr-β-TCP (**f**) and Mg-β-TCP (**g**) composite coatings.

An AFM study of thin RFMS surface layer showed the relationship between surface structure and composition of the sputtered target. We observed RFMS crystallites over the PEO spherulites surface in the images of composite coatings (Figure 2b–g). Homogeneously distributed crystallites are clearly visible in the images of PEO + HA (Figure 2(b2)) and PEO + Mg-HA (Figure 2(d2)). HA crystallites are oval and have a 0.31 ± 0.06 µm^2 approximate area (Figure 2(b3)). Mg-HA crystallites are smaller and near the quasi-equilibrium shape with an approximate area of 0.06 ± 0.02 µm^2 (Figure 2(d3)). However, most of the composite coatings, including PEO + Sr-HA (Figure 2(c1–c3)), PEO + β-TCP (Figure 2(e1–e3)), PEO + Sr-β-TCP (Figure 2(f1–f3)), and PEO + Mg-β-TCP (Figure 2(g1–g3)), are formed by HA-type crystallites. Presumably, Figure 2(c3,f3) shows initial PEO spherulites with no traces of RFMS crystallites.

3.2. Coating Phase Composition

In the XRD patterns of all composite coatings, we observed metallic substrate α-Ti characteristic peaks at 2θ = 35.24° (corresponding to 100 plane reflection), 38.39° (002), 40.29° (101), 63.24° (110), 74.52° (200), 76.47° (112) and 77.71° (201) (Figure 3). In the patterns of HA and Sr-HA RFMS coatings, we observed sharp peaks in hydroxyapatite $Ca_5(PO_4)_3(OH)$ at 2θ = 25.85° (002), 31.77° (121), 32.18° (112), 32.91° (300), 34.05° (202), 46.69° (222) and 49.46° (213) (Figure 3a). We also identified less intense hydroxyapatite peaks at 2θ = 28.11° (102), 28.93° (210), 52.07° (402), 53.09° (141) and 70.79° (503). Considering the absence of the reflection from the (211) plane, which is the strongest line of HA phase and the width and sharpness of the observed peaks, we suggest that composite coatings' RFMS with HA and Sr-HA are predominantly amorphous coatings, with the presence of the peaks in the HA crystal phase.

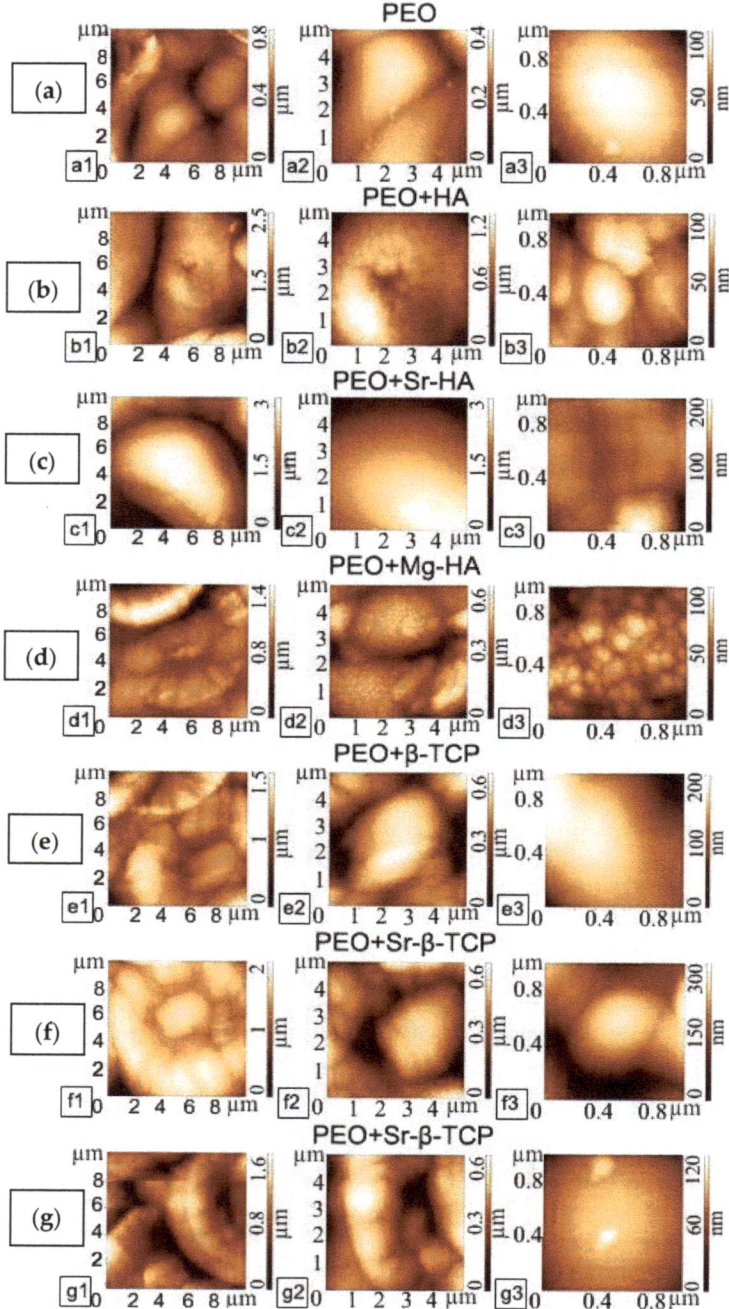

Figure 2. AFM images of CaP PEO (**a**) and composite PEO + RF coatings sputtered with HA (**b**), Sr-HA (**c**), Mg-HA (**d**), β-TCP (**e**), Sr-β-TCP (**f**), and Mg-β-TCP (**g**) targets.

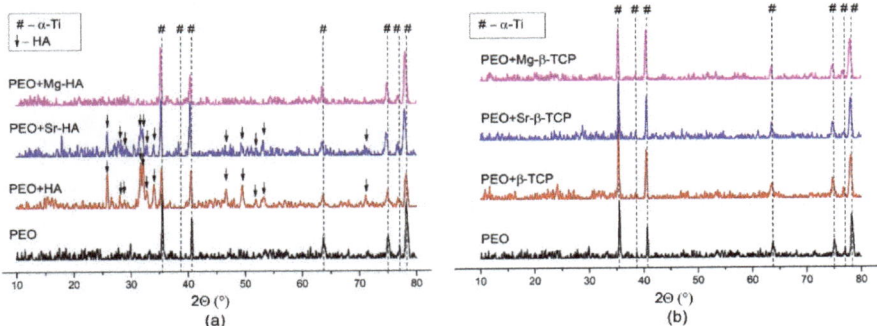

Figure 3. XRD patterns of CaP PEO and composite PEO + RF coatings sputtered with (**a**) HA, Sr-HA, Mg-HA; (**b**) β-TCP, Sr-β-TCP, Mg-β-TCP targets.

3.3. Coating Chemical Composition

Chemical composition of composite coatings is presented in Table 2. The compositions of coatings correspond to the chemical composition of PEO electrolyte and RFMS targets and are generally represented by Ca, P and O. Coatings' RFMS with Sr- and Mg-substituted targets contain Sr and Mg, respectively.

Table 2. Chemical composition of CaP composite coatings.

Sample	The Content of Elements in the Coating (at.%)							
	O	Ca	P	Ti	C	Sr	Mg	Ca/P
PEO	61.5 ± 0.6	6.0 ± 0.3	17.6 ± 0.2	3.7 ± 0.3	11.4 ± 1.4	–	–	0.34 ± 0.02
PEO + HA	54.9 ± 0.8 *	17.9 ± 0.6 *	14.7 ± 0.5 *	–	12.5 ± 0.5	–	–	1.22 ± 0.06 *
PEO + Sr-HA	57.7 ± 1.8 *	17.1 ± 0.5 *	17.0 ± 0.5	–	7.7 ± 2.8	0.5 ± 0.04 *	–	1.01 ± 0.04 *
PEO + Mg-HA	60.0 ± 2.0	13.4 ± 0.4 *	17.4 ± 0.6	–	7.9 ± 2.9	–	1.3 ± 0.3 *	0.77 ± 0.04 *
PEO + β-TCP	59.4 ± 1.0 *†	15.1 ± 0.1 *†	16.2 ± 0.4 *†	–	9.3 ± 1.4 †	–	–	0.93 ± 0.02 *†
PEO + Sr-β-TCP	58.3 ± 2.5	13.2 ± 0.8 *†	19.0 ± 1.3 *†	–	9.0 ± 4.7	0.5 ± 0.2 *	–	0.70 ± 0.06 *†
PEO + Mg-β-TCP	62.8 ± 0.5 *†	8.5 ± 0.1 *†	18.3 ± 0.3 *	–	7.4 ± 1.2 *	–	2.9 ± 0.4 *†	0.46 ± 0.01 *†

* significant differences with PEO ($p < 0.05$); † significant differences between groups of the X-β-TCP and X-HA coatings, where X is Mg or Sr ($p < 0.05$).

It should be noted that CaP PEO coatings are generally calcium-deficient and characterized by Ca/P = 0.34 ± 0.02. The RFMS upper coating increases the Ca/P ratio. The Ca/P ratio varies in the following progression: PEO < PEO + Mg-β-TCP < PEO + Sr-β-TCP < PEO + Mg-HA < PEO + β-TCP < PEO + Sr-HA < PEO + HA. Moreover, for both HA and β-TCP, the Ca/P ratio is greater for Sr-substituted upper coatings compared with Mg-substituted coatings.

In the XPS spectra of PEO coating, we observed typical O1s, Ti2p, Ca2p, C1s, P2p and Auger's peaks corresponding to O KLL and Ca LMM (Figure 4). The Ti2p peak appears only in PEO CaP coatings' spectra at the following binding energy: Ti2p3/2 = 459.28 eV. Generally, this peak recognizes O–Ti–O of TiO$_2$ [23]. In composite coatings' spectra, Ti2p is not identified due to the RFMS thin CaP upper coating. In the spectra of composite coatings sputtered with Sr- and Mg-substituted HA and β-TCP, we revealed the small peaks in Sr and Mg, respectively. The typical XPS spectra of Ca2p and P2p core levels of the coatings are shown in Figure 5. The corresponding BE values of Ca2p3/2 and P2p3/2 of the coatings are listed in Table 3. Generally, different CaP materials show peaks at similar binding energies in the XPS spectra [24,25]. Peaks overlapping makes it difficult to distinguish different CaP phases in composite material. Peaks at 347.46, 133.00 and 347.46, 133.00 eV in the spectra of PEO + β-TCP and PEO + Sr-β-TCP, respectively, correspond to Ca 2p3/2 and P 2p3/2 in Ca$_5$(PO$_4$)(OH).

This correlates with the XRD data. XPS analysis revealed the following relations for composite coatings (Table 3):

- PEO and PEO + Mg-β-TCP coatings show the highest P2*p* binding energy;
- PEO + Mg-β-TCP and PEO+Mg-HA coatings show the highest Ca2*p* binding energy;
- Binding energy increases by the following progression: PEO + Mg-β-TCP < PEO + Sr-β-TCP < PEO + β-TCP and PEO + Mg-HA < PEO + Sr-HA < PEO + HA.

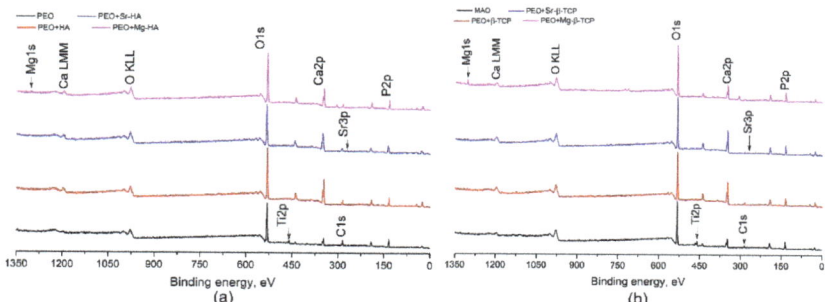

Figure 4. The XPS survey spectra for the CaP PEO (black line on Figure 4a,b) and composite HA (**a**) and β-TCP (**b**) RFMS coatings.

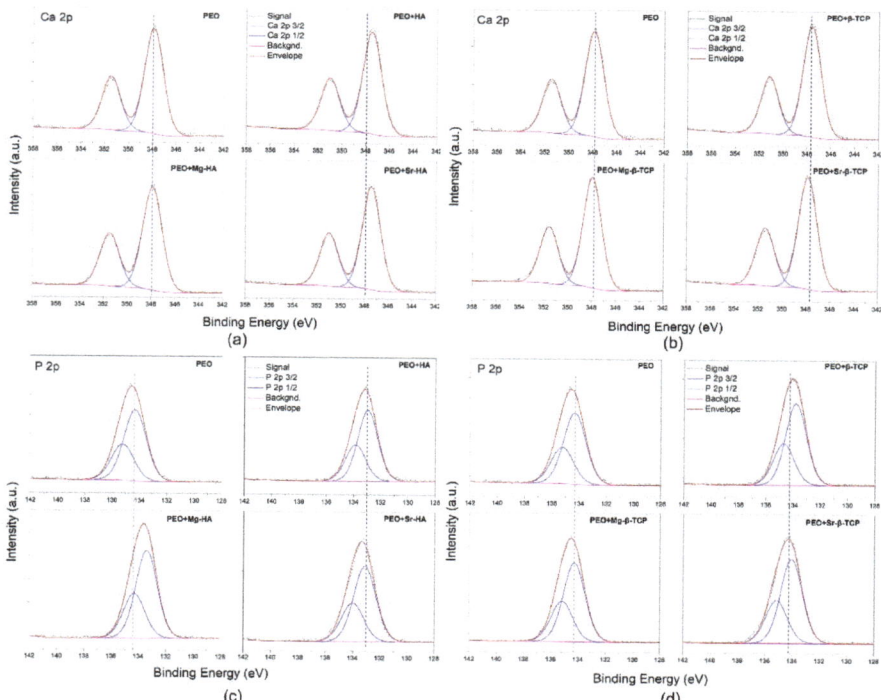

Figure 5. High-resolution XPS spectra of Ca 2*p* (**a**,**b**) and P 2*p* (**c**,**d**) core levels of composite CaP coatings.

Table 3. Binding energy values of the Ca2p3/2 and P2p3/2 XPS peaks.

Coating	Ca 2p3/2 (eV)	P 2p3/2 (eV)
PEO	347.88	134.31
PEO + HA	347.46	133.00
PEO + Sr-HA	347.44	133.10
PEO + Mg-HA	347.92	133.43
PEO + β-TCP	347.59	133.80
PEO + Sr-β-TCP	347.87	134.07
PEO + Mg-β-TCP	347.97	134.31

3.4. Young's Modulus

RFMS CaP upper coating significantly enhances PEO + Mg-HA, PEO + Sr-HA and PEO + β-TCP Young's modulus compared with the PEO subcoating (Table 4). It is important to note that the Young's modulus of all coatings under study varies in the range 15–37 GPa, which corresponds with the nanoindentation modulus of human bone [26].

Table 4. Young's modulus of CaP PEO and composite coatings.

Coating	E_{IT}, GPa
PEO	15.4 ± 3.8
PEO + HA	21.3 ± 3.7
PEO + Sr-HA	36.6 ± 9.5 *,†
PEO + Mg-HA	29.8 ± 6.8 *
PEO + β-TCP	25.6 ± 5.8 *
PEO + Sr-β-TCP	19.5 ± 4.1
PEO + Mg-β-TCP	17.4 ± 3.2

* significant differences with PEO, $p < 0.05$; † significant differences between groups of the β-TCP and X-β-TCP coatings and HA and X-HA coatings, where X is Mg or Sr, $p < 0.05$.

4. Conclusions

Composite PEO+RFMS coatings were characterized by the multileveled roughness, increased Ca/P ratio and Young's modulus (PEO + Mg-HA, PEO + Sr-HA, PEO + β-TCP). The upper RFMS CaP coating forms a thin layer over the PEO sub-coating spherulites. The RFMS of different CaP and ion-substituted CaP provides various coating crystallinities and are potentially beneficial in adjustable biodegradation and osseointegration. Moreover, the sputtering of ion-substituted CaP targets makes it possible to introduce into the composition of the coatings elements that bone tissue growth-stimulating (Mg) and suppress the resorptive activity of osteoclasts (Sr).

Author Contributions: A.K.: conceptualization, data curation, funding acquisition, supervision, investigation, methodology, validation, writing—original draft, writing—review and editing; G.D.: nanoindentation, data curation; validation; writing—review and editing; A.V.: PEO subcoating formation, data curation; validation; A.F.: upper coatings' formation by RFMS, data curation; validation; N.K.: synthesizing the powders of pure hydroxyapatite (HA), Sr-HA, Mg-HA by microwave-assisted method, data curation; validation; A.G.: synthesizing the powders of pure hydroxyapatite (HA), Sr-HA, Mg-HA by microwave-assisted method, data curation; validation; E.S.: upper coatings' formation by RFMS, data curation; validation; Y.Z.: X-ray photoelectron spectroscopy (XPS), data curation; validation; S.T.: conceptualization, resources, funding acquisition, supervision, validation. All authors have read and agreed to the published version of the manuscript.

Funding: This work was supported by the Tomsk Polytechnic University Competitiveness Enhancement Program project VIU-School of Nuclear Science and Engineering-204/2019 and project VIU-SEC B.P. Veinberg-196/2020.

Acknowledgments: Our gratitude to Liga Stipniece and Janis Locs from Riga Technical University (Riga, Latvia) for the synthesis of the β-TCP, Mg-β-TCP, Sr-β-TCP powders. We are grateful to V.P. Ignatov for consultations on the preparation of CaP electrolyte for PEO and valuable advice on the discussion of the experimental results. The authors are grateful to the Resource Centre "Materials Science Shared Center", part of the "Tomsk Regional Common Use Center (TRCUC)" of Tomsk State University, for XRD measurements, and the Resource Centre "Physical Methods of Surface Investigation" of Saint-Petersburg State University Research Park for XPS measurements.

Conflicts of Interest: The authors declare no conflict of interest.

References

1. Liu, F.; Song, Y.; Wang, F.; Shimizu, T.; Igarashi, K.; Zhao, L. Formation characterization of hydroxyapatite on titanium by microarc oxidation and hydrothermal treatment. *J. Biosci. Bioeng.* **2005**, *100*, 100–104. [CrossRef] [PubMed]
2. Parfenov, E.; Parfenova, L.; Mukaeva, V.; Farrakhov, R.; Stotskiy, A.; Raab, A.; Danilko, K.; Rameshbabu, N.; Valiev, R. Biofunctionalization of PEO coatings on titanium implants with inorganic and organic substances. *Surf. Coat. Technol.* **2020**, *404*, 126486. [CrossRef]
3. Yang, W.; Wang, P.; Guo, Y.; Jiang, B.; Yang, F.; Li, J. Microstructure and corrosion resistance of modified AZ31 magnesium alloy using microarc oxidation combined with electrophoresis process. *J. Wuhan Univ. Technol. Sci. Ed.* **2013**, *28*, 612–616. [CrossRef]
4. Rehman, Z.U.; Choi, D.; Koo, B.H. A three layer corrosion protection system for Mg–9Al–Zn alloy offered by complex fluoride ions and precipitates-based electrolytic oxidation. *Surf. Coat. Technol.* **2020**, *393*, 125804. [CrossRef]
5. Ur Rehman, Z.; Heun Koo, B.; Choi, D. Influence of complex SiF_6^{2-} Ions on the PEO coatings formed on Mg–Al6–Zn1 alloy for enhanced wear and corrosion protection. *Coatings* **2020**, *10*, 94. [CrossRef]
6. Ur Rehman, Z.; Choi, D. Investigation of ZrO_2 nanoparticles concentration and processing time effect on the localized PEO coatings formed on AZ91 alloy. *J. Magnes. Alloy.* **2019**, *7*, 555–565. [CrossRef]
7. Gu, Y.; Xiong, W.; Ning, C.; Zhang, J. Residual stresses in microarc oxidation ceramic coatings on biocompatible AZ31 magnesium alloys. *J. Mater. Eng. Perform.* **2011**. [CrossRef]
8. Akin, F.A.; Zreiqat, H.; Jordan, S.; Wijesundara, M.B.J.; Hanley, L. Preparation and analysis of macroporous TiO_2 films on Ti surfaces for bone–tissue implants. *J. Biomed. Mater. Res. Part A* **2001**, *57*, 588–596. [CrossRef]
9. Yavari, S.A.; Necula, B.S.; Fratila-Apachitei, L.E.; Duszczyk, J.; Apachitei, I. Biofunctional surfaces by plasma electrolytic oxidation on titanium biomedical alloys. *Surf. Eng.* **2016**, *32*, 411–417. [CrossRef]
10. Krząkała, A.; Kazek-Kęsik, A.; Simka, W. Application of plasma electrolytic oxidation to bioactive surface formation on titanium and its alloys. *RSC Adv.* **2013**, *3*, 19725. [CrossRef]
11. Barati Darband, G.; Aliofkhazraei, M.; Hamghalam, P.; Valizade, N. Plasma electrolytic oxidation of magnesium and its alloys: Mechanism, properties and applications. *J. Magnes. Alloy.* **2017**, *5*, 74–132. [CrossRef]
12. Jiang, B.L.; Wang, Y.M. Plasma electrolytic oxidation treatment of aluminium and titanium alloys. In *Surface Engineering of Light Alloys*; Elsevier: Amsterdam, The Netherlands, 2010; pp. 110–154.
13. Depla, D.; Mahieu, S.; Greene, J.E. Sputter Deposition Processes. In *B Handbook of Deposition Technologies for Films and Coatings*; Elsevier: Amsterdam, The Netherlands, 2010; pp. 253–296.
14. Juhasz, J.A.; Best, S.M. Surface modification of biomaterials by calcium phosphate deposition. In *Surface Modification of Biomaterials*; Elsevier: Amsterdam, The Netherlands, 2011; pp. 143–169.
15. Park, S.-Y.; Choe, H.-C. Mn-coatings on the micro-pore formed Ti–29Nb–xHf alloys by RF-magnetron sputtering for dental applications. *Appl. Surf. Sci.* **2018**, *432*, 278–284. [CrossRef]
16. Hwang, I.-J.; Choe, H.-C. Surface morphology and cell behavior of Zn-coated Ti–6Al–4V alloy by RF-sputtering after PEO-treatment. *Surf. Coat. Technol.* **2019**, *361*, 386–395. [CrossRef]
17. Lu, J.; Yu, H.; Chen, C. Biological properties of calcium phosphate biomaterials for bone repair: A review. *RSC Adv.* **2018**, *8*, 2015–2033. [CrossRef]
18. Urquia Edreira, E.R.; Wolke, J.G.C.; Aldosari, A.A.; Al-Johany, S.S.; Anil, S.; Jansen, J.A.; van den Beucken, J.J.J.P. Effects of calcium phosphate composition in sputter coatings on in vitro and in vivo performance. *J. Biomed. Mater. Res. Part A* **2015**, *103*, 300–310. [CrossRef]
19. Eliaz, N.; Metoki, N. Calcium phosphate bioceramics: A review of their history, structure, properties, coating technologies and biomedical applications. *Materials* **2017**, *10*, 334. [CrossRef]
20. Zeng, J.; Guo, J.; Sun, Z.; Deng, F.; Ning, C.; Xie, Y. Osteoblastic and anti-osteoclastic activities of strontium-substituted silicocarnotite ceramics: In vitro and in vivo studies. *Bioact. Mater.* **2020**, *5*, 435–446. [CrossRef]

21. BS EN ISO 14577 International Organization for Standardization. *Metallic Materials: Instrumented Indentation Test for Hardness and Materials Parameters. Verification and Calibration of Testing Machines*; ISO: Geneva, Switzerland, 2002.
22. Kozelskaya, A.I.; Kulkova, S.E.; Fedotkin, A.Y.; Bolbasov, E.N.; Zhukov, Y.M.; Stipniece, L.; Bakulin, A.V.; Useinov, A.S.; Shesterikov, E.V.; Locs, J.; et al. Radio frequency magnetron sputtering of Sr- and Mg-substituted β-tricalcium phosphate: Analysis of the physicochemical properties and deposition rate of coatings. *Appl. Surf. Sci.* **2020**, *509*, 144763. [CrossRef]
23. Callen, B.W.; Lowenberg, B.F.; Lugowski, S.; Sodhi, R.N.S.; Davies, J.E. Nitric acid passivation of Ti6A14V reduces thickness of surface oxide layer and increases trace element release. *J. Biomed. Mater. Res.* **1995**, *29*, 279–290. [CrossRef]
24. França, R.; Samani, T.D.; Bayade, G.; Yahia, L.; Sacher, E. Nanoscale surface characterization of biphasic calcium phosphate, with comparisons to calcium hydroxyapatite and β-tricalcium phosphate bioceramics. *J. Colloid Interface Sci.* **2014**, *420*, 182–188. [CrossRef]
25. Lu, H.B.; Campbell, C.T.; Graham, D.J.; Ratner, B.D. Surface characterization of hydroxyapatite and related calcium phosphates by XPS and TOF-SIMS. *Anal. Chem.* **2000**, *72*, 2886–2894. [CrossRef]
26. Wu, D.; Isaksson, P.; Ferguson, S.J.; Persson, C. Young's modulus of trabecular bone at the tissue level: A review. *Acta Biomater.* **2018**, *78*, 1–12. [CrossRef]

Publisher's Note: MDPI stays neutral with regard to jurisdictional claims in published maps and institutional affiliations.

© 2020 by the authors. Licensee MDPI, Basel, Switzerland. This article is an open access article distributed under the terms and conditions of the Creative Commons Attribution (CC BY) license (http://creativecommons.org/licenses/by/4.0/).

Article

Characterization, Bioactivity and Antibacterial Properties of Copper-Based TiO$_2$ Bioceramic Coatings Fabricated on Titanium

Salih Durdu

Department of Industrial Engineering, Giresun University, Giresun 28200, Turkey; durdusalih@gmail.com or salih.durdu@giresun.edu.tr; Tel.: +90-4543104114

Received: 15 October 2018; Accepted: 18 December 2018; Published: 20 December 2018

Abstract: The bioactive and anti-bacterial Cu-based bioceramic TiO$_2$ coatings have been fabricated on cp-Ti (Grade 2) by two-steps. These two-steps combine micro-arc oxidation (MAO) and physical vapor deposition–thermal evaporation (PVD-TE) techniques for dental implant applications. As a first step, all surfaces of cp-Ti substrate were coated by MAO technique in an alkaline electrolyte, consisting of Na$_3$PO$_4$ and KOH in de-ionized water. Then, as a second step, a copper (Cu) nano-layer with 5 nm thickness was deposited on the MAO by PVD-TE technique. Phase structure, morphology, elemental amounts, thickness, roughness and wettability of the MAO and Cu-based MAO coating surfaces were characterized by XRD (powder- and TF-XRD), SEM, EDS, eddy current device, surface profilometer and contact angle goniometer, respectively. The powder- and TF-XRD spectral analyses showed that Ti, TiO$_2$, anatase-TiO$_2$ and rutile-TiO$_2$ existed on the MAO and Cu-based MAO coatings' surfaces. All coatings' surfaces were porous and rough, owing to the presence of micro sparks through MAO. Furthermore, the surface morphology of Cu-based MAO was not changed. Also, the Cu-based MAO coating has more hydrophilic properties than the MAO coating. In vitro bioactivity and in vitro antibacterial properties of the coatings have been investigated by immersion in simulated body fluid (SBF) at 36.5 °C for 28 days and bacterial adhesion for gram-positive (*S. aureus*) and gram-negative (*E. coli*) bacteria, respectively. The apatite layer was formed on the MAO and Cu-based MAO surfaces at post-immersion in SBF and therefore, the bioactivity of Cu-based MAO surface was increased to the MAO surface. Also, for *S. aureus* and *E. coli*, the antibacterial properties of Cu-based MAO coatings were significantly improved compared to one of the uncoated MAO surfaces. These results suggested that Cu-based MAO coatings on cp-Ti could be a promising candidate for biomedical dental implant applications.

Keywords: micro arc oxidation (MAO); Cu nano-layer; hydrophilic surface; apatite; in vitro bioactivity; antibacterial properties

1. Introduction

Commercially pure titanium (cp-Ti) materials are preferred for dental implant applications, owing to its low density, low elastic modulus (closer to that of bone), low thermal conductivity, non-magnetic properties, high specific strength, corrosion resistance, good mechanical properties, fracture resistance and fatigue resistance and biocompatibility [1–3]. It is well-known that titanium has corrosion resistant and biocompatibility properties. These are related to the native TiO$_2$ layer spontaneously formed on its surface [4,5]. However, titanium implant materials cannot bond directly to the bone owing to their bio-inert nature (not bioactive), unlike bioactive ceramics such as bio-glass, glass ceramic, hydroxyapatite (HA), ZrO$_2$ and TiO$_2$ etc. [6,7]. As a result of this, the bone tissue around the implant is absorbed. This leads to slow healing and the loosening of the implant–bone interface [8]. Therefore, bioactive ceramics such as HA or TiO$_2$ on titanium were coated to enhance bioactivity [9,10].

TiO$_2$-based coatings have been suggested to improve corrosion resistance, bioactivity and biocompatibility of implant surfaces. They have recently received great attention for the biomedical applications owing to their more stable chemical composition [11,12]. TiO$_2$ were fabricated on cp-Ti by various surface coating methods including sol–gel [13], anodic oxidation [14], magnetron sputtering [15], electrophoretic deposition [16], acid etching [17,18], laser surface treatment [19], plasma spraying [20,21] and micro arc oxidation [22–25] etc. However, certain problems like non-homogenous structure, micro-structural control problems, micro-cracks formation, the presence of phase impurity, poor adhesion strength were observed in these methods except for MAO technique [26–28]. Thus, implant application areas of the coatings produced by these techniques are limited.

Micro-arc oxidation (MAO) is one of the most applicable methods to deposit a porous and rough bioceramic layer on valve metals such as Ti, Al, Mg, and Zr surfaces [3,29–31]. The MAO coating promotes bioactivity, biocompatibility, wear and corrosion resistance respect to other surface coating methods [32–39]. Also, the adhesion strength between the substrate/MAO coatings is excellent, owing to its in situ growth [40]. Furthermore, it is reported that the porous and rough surfaces on the implant surfaces are beneficial for the formation and Osseo-integration of new bone tissue [41]. However, the bacterial adhesion and colonization may occur on the implant surfaces under body conditions. These lead to infections at the implant site and it results in the loss of implants [42].

In order to overcome bacterial adhesion and colonization, the surface modifications of implants were increasingly carried out by numerous anti-bacterial agents such as Cu [12], Ag [43] and Zn [44] over the last few years. The Cu is one of the basic trace elements necessary for human existence. It contributes in synthesis and release of life-sustaining proteins and enzymes within the living organisms [45]. Furthermore, it actively takes part in various enzyme-based processes, for bone metabolism stimulates new vessel formation and accelerates early skin wound healing [46,47]. Therefore, it is beneficial for bone tissue formation [12]. Moreover, copper exhibits excellent antibacterial properties versus a broad spectrum of bacteria, including gram-positive and gram-negative bacteria by interfering DNA replication and disrupting cell membranes [48–50].

In particular, there have been some promising studies on the fabrication of antibacterial Cu/CuO-nanoparticles containing Cu-incorporated TiO$_2$ coatings on cp-Ti by using the MAO technique for the last five years [42,45,51–56]. Wu et al. investigated the formation and investigation of antibacterial resistances of Cu-incorporated TiO$_2$ coatings by MAO and hydrothermal treatment (HT) [42]. Yao et al. investigated the antibacterial properties of Cu nanoparticles containing TiO$_2$ coating synthesized by MAO [45]. Huang et al. fabricated the Cu-incorporated bioceramic coatings by MAO and HT and they evaluated osteoblast response [51]. Zhu et al. produced Cu-containing micro arc oxidized TiO$_2$ coatings and evaluated anti-bacterial properties [52]. Zhang et al. examined the antibacterial properties of TiO$_2$ coatings doped with various amounts with Cu nanoparticles deposited on titanium by MAO [53]. Huang et al. prepared Cu-containing TiO$_2$ coatings on Ti by MAO and then, biocompatibility and antibacterial properties were evaluated [54]. Zhang et al. directly fabricated Cu-doped TiO$_2$ coatings in alkaline electrolyte containing β-glycerophosphate disodium, calcium acetate and various amounts of copper acetate on Ti via a single step MAO process [55]. They then investigated in vitro biocompatibility and antibacterial activity for gram positive *S. aureus* [55]. However, in the above studies, Cu nanoparticles were randomly separated through the whole MAO surface. Also, for Cu-incorporated- and Cu containing-MAO coatings, Cu was not homogeneously distributed throughout the MAO surface. So, antibacterial adhesion properties on the Cu-MAO surface could not be sufficiently improved. Furthermore, Wu et al. produced Cu-doped TiO$_2$ coatings on cp-Ti by magnetron sputtering with MAO [56]. The CuTi layers were formed on the titanium substrate by magnetron sputtering at the first stage. Then, the bioceramic coatings were produced by MAO technique at the second stage. Afterwards, their antibacterial properties were evaluated [56]. However, in that study, the Cu was observed around micro discharge channels and could not be dispersed throughout the MAO surface. Moreover, in vitro bioactivity of the Cu-MAO coatings were not investigated, although in vitro biocompatibility investigations were carried out in the above aforementioned studies.

In this work, unlike the literature, convenient two-step MAO and PVD-TE techniques were devoted to synthesis uniform, bioactive, biocompatible and anti-bacterial novel Cu-based TiO_2 bioceramic composite coatings on cp-Ti substrate. A bioactive and biocompatible anatase and rutile-based bioceramic structure on the cp-Ti substrate were coated by MAO technique in an alkaline electrolyte, consisting of Na_3PO_4 and KOH in de-ionized water at the first step. Then, a copper (Cu) nano-layer with 5 nm thickness was accumulated on the MAO coatings by the PVD-TE technique at the second step. The phase structures, morphologies, elemental amounts, functional groups, thicknesses, roughness and wettability of the MAO and Cu-based MAO coating surfaces were characterized by XRD (powder- and TF-XRD), SEM, EDS, FTIR, eddy current device, profilometer and CAG in detail, respectively. In vitro bioactivity of all coatings was evaluated by immersion tests in SBF at body temperature (36.5 °C) for 28 days. Then, for gram-positive bacteria (*Staphylococcus aureus*) and gram-negative bacteria (*Escherichia coli*), in vitro antibacterial properties of all coatings were investigated in detail.

2. Materials and Methods

2.1. Sample Preparation

The cp-Ti (Grade 2; commercially pure titanium) substrates were cut appropriate sizes (60 mm × 25 mm × 5 mm) by using a water jet. The surfaces of cp-Ti were ground by using silicon carbide (SiC) sand papers from No. 120 to No. 1200. And then, they were cleaned in an ultrasonic bath containing acetone for 60 min and dried in an oven at 50 °C.

2.2. Micro Arc Oxidation (MAO) Process

In order to produce bioceramic coatings on cp-Ti, an alternating current (AC) MAO device (MDO-100WS) running up to 100 kW was used. The MAO device mainly contained four pieces of equipment, consisting of an AC power supply, a double walled stainless steel tank, water cooled chiller and air flow stirrer. The cp-Ti was used as an anode, while the stainless-steel container was used as a cathode during MAO. The MAO solution was prepared by the dissociation of 10 g/L Na_3PO_4 and 1 g/L KOH in de-ionized water, respectively. The MAO treatment was performed on a constant current mode in the range of 0.325 A/cm^2 in Na_3PO_4 and KOH. The treating time was carried out at 5 min. The detailed MAO parameters and analysis results are given in Table 1. The temperature was maintained below 30 °C by a chiller in the tank during the MAO treatment. After the MAO treatment, all substrates were rinsed by de-ionized water and dried again in an oven at 50 °C for 24 h. Afterwards, they were preserved in desiccators. In order to ensure repeatability during MAO process, three MAO surfaces were produced on three cp-Ti specimens by the same parameters.

Table 1. The MAO coating parameters and the surface analysis results of the coatings such as thickness.

Electrolyte	Treatment Time (min)	Applied Power (µF)	The Sizes (mm^3)	Current Density (A/cm^2)	Average Thickness (µm)	Average Roughness (µm)
Na_3PO_4, KOH, Distilled water	5	150	60 × 25 × 5	0.325	60.0 ± 1.0	1.07

2.3. Physical Vapor Deposition-Thermal Evaporation (PVD-TE) Process

A Cu thin film layer with 5 nm (Copper: 99.999% purity of Alfa Aesar, Ward Hill, MA, USA) was accumulated on the MAO coatings using PVD-TE (Vaksis, Bilkent, Turkey, PVD/2T) at a deposition speed of 0.5 nm/s at room temperature. The vacuum chamber pressure was set at about 1×10^{-6} mbar before the PVD-TE process started. The vacuum chamber pressure was maintained under vacuum (base pressure) of 3×10^{-5} mbar. In order to avoid the abrupt evaporation of Cu powders, the changeable current was gradually increased up to 45 A. The evaporated materials (Cu powders) with a grain size of -100 mesh were placed at the bottom of a wolfram crucible, that was approximately 15 cm away from the MAO coating surfaces. The average thickness of Cu nano-layer on the MAO surfaces was measured as about 5 nm by XTM integrated to PVD-TE device (Vaksis, Bilkent, Turkey, PVD/2T). Cu vapor products were then deposited onto the MAO surfaces. In order to ensure repeatability during the PVD-TE process, three Cu nano-layers were deposited on three MAO coating specimens by the same PVD-TE parameters.

2.4. Characterization of the Coatings

The phase structure of the MAO surfaces was detected by powder-XRD (powder-X-ray diffractometer, Bruker D8 Advance, Billerica, MA, USA) with Cu-Kα (λ = 1.54 Å) between 2θ values of 10° and 90° with a scanning rate of $0.1° \cdot min^{-1}$. The phase structure of Cu-based MAO coatings was analyzed by TF-XRD (Thin film-X-ray diffractometer, PANalytical X'Pert PRO MPD, Philips, Amsterdam, The Netherlands) with Cu-Kα between 2θ values of 10° and 90° with a scanning rate of $0.001° \cdot min^{-1}$. The surface morphologies of all coatings were analyzed by SEM (Scanning electron microscope, Philips XL30S FEG, Amsterdam, The Netherlands). Also, the elemental amounts of all coatings were evaluated by EDS (Energy dispersive spectrometer, Philips, Amsterdam, The Netherlands). The 3-D surface topography and surface roughness were evaluated by profilometer (surface profiler, KLA Tencor P-7, Milpitas, CA, USA). The surface roughness values were achieved by the scanning of mechanical contact at 500 μm × 500 μm area in 3-D. The surface wettability (hydrophilicy/hydrophobicity) and the contact angle values of all surfaces were analyzed by CAG (Contact Angle Goniometer, Dataphysics OCA 15EC, San Jose, CA, USA) sessile drop technique. The contact angle data was recorded throughout every 10 s from 0 to 90 s after the de-ionized water droplet of volume 1 μL was contacted onto the both surfaces.

2.5. In Vitro Bioactivity Tests

To be informed of in vitro apatite-forming ability on the MAO and Cu-based MAO surfaces, all coatings were immersed in 1.0× SBF (simulated body fluid) at body temperature (36.5 °C) for 28 days. The SBF procures the formation of bone-like apatite layer on the implant surfaces. So, the apatite-forming ability on the implant materials represents information predicted about in vitro bioactivity. The MAO and Cu-based MAO coating samples were immersed in 1.0× SBF with ion concentrations almost equal to that in human blood plasma. The SBF was prepared by dissolving reagent grade chemicals consisting of NaCl, $NaHCO_3$, KCl, $K_2HPO_4 \cdot 3H_2O$, $MgCl_2 \cdot 6H_2O$, $CaCl_2$ and Na_2SO_4 in distilled water at 36.5 °C, respectively. They were then buffered at pH 7.4 with $(CH_2OH)_3CNH_2$ and 1 M HCl at 36.5 °C. The surface area' ratio (in mm^2) of all surfaces to SBF was almost set 1 equal to 10 in the direction of the Kokubo and Takadama' recipe [57]. To maintain ion concentration of the SBF, it was renewed during every 24 h. All coating specimens were taken out from SBF at the post-immersion test and they were gently washed in de-ionized water. Eventually, they spontaneously dried under room temperature. All immersed dried coating specimens were kept in desiccators at pre-characterization. All experimental studies were carried out in triplicate.

After immersion treatment was completed in SBF, the morphologies, elemental structures, phase structures and functional groups of all coating surfaces were analyzed by SEM, EDS, TF-XRD and FTIR, respectively. The SEM images were taken with up to 20000× magnification. In addition, to reveal newly formed elements on immersed surfaces, all coatings surfaces were examined by EDS. The phase compositions of both coating surfaces were investigated by TF-XRD. The FTIR (Fourier transform infrared spectroscopy; JASCO FT/IR 6600, JASCO, Easton, MD, USA) spectra were collected over the range in the spectral range of 450–4000 cm^{-1} at post-immersion in SBF.

2.6. In Vitro Antibacterial Activity of the Coatings

The antibacterial activities of all coating surfaces were investigated versus to *S. aureus* and *E. coli* by colony counting method. All coating surfaces were immersed in 5.0 mL of the bacterial suspension (1×10^7 CFU/mL). They were then incubated at 37 °C for 24 h. All coatings samples were washed by 150 mM NaCl at post-incubation and put into a tube including 2 mL phosphate buffer solution. Subsequently, to detach the bacteria from the surfaces to solution, they were shaken on a vortex for 2 min. Aliquots of the solution with 100 µL were plated onto muller hinton agar (MHA) plates. The active bacteria colonies on the surfaces were then incubated at 37 °C for 48 h, and were counted. All experimental studies were carried out in triplicate.

3. Results and Discussion

3.1. Phase Structures of the Coatings

The phase structures of the MAO and Cu-based MAO coatings on cp-Ti were investigated by powder XRD and TF-XRD as shown in Figures 1 and 2, respectively. In addition to the cp-Ti substrate diffraction peaks (JCPDS card number: 044-1294), the existences of characteristic peaks of anatase-TiO_2 (JCPDS card number: 21-1272) and rutile-TiO_2 (JCPDS card number: 21-1276) on the MAO surface were indicated by the powder-XRD pattern as shown in Figure 1.

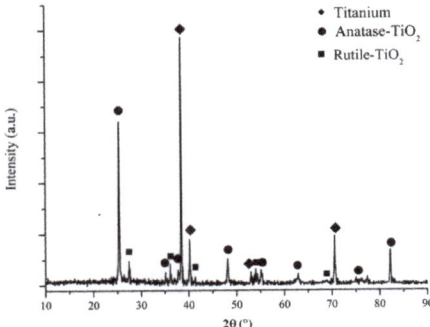

Figure 1. XRD spectra pattern of the MAO coating.

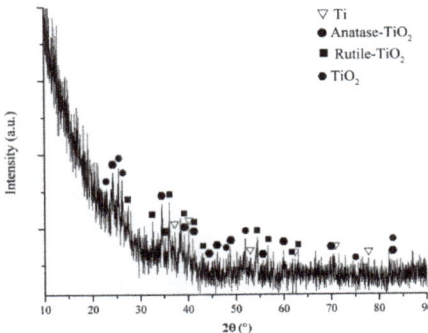

Figure 2. TF-XRD spectra pattern of the Cu-based MAO coating.

After Cu was accumulated on the MAO surface, the phase structures formed on the surface were detected by TF-XRD as shown in Figure 2. The Ti (JCPDS card number: 04-004-8480), TiO_2 (JCPDS card number: 01-070-2556), anatase-TiO_2 (JCPDS card number: 01-083-5914) and rutile-TiO_2 (JCPDS card number: 04-006-8034) were obtained on the Cu-based MAO surface as illustrated in Figure 1. The presence of crystalline Cu and/or Cu-based compounds were not verified by TF-XRD.

The phase of TiO_2 has two polymorphs, which are also known as anatase and rutile. The phase of rutile is stable at high temperatures, while the anatase is a metastable phase at low temperature. The phase of anatase forms, which is a metastable structure, on cp-Ti surface at initial steps of MAO process. It is reported that the amount of anatase in the coatings decreases, as the amount of rutile increases. It is clearly stated that the rutile modifier becomes predominant after a critical period of MAO parameters such as treatment time, voltage and current. So, the anatase transforms to thermodynamically stable rutile with increasing experimental parameters, such as treatment time, voltage and current at the next stages of the MAO process [9,58].

Anatase and rutile phases formed on the cp-Ti surface through the MAO process occurred by the ionization mechanism and electrostatic interactions of oppositely charged ions (anion and cation). The alkaline electrolyte consists of Na_3PO_4, and KOH compounds contain Na^+, PO_4^{3-}, K^+ and OH^- ions. The Na_3PO_4 is dissolved in distilled water and ionizes to Na^+ and PO_4^{3-} (Equation (1)). Similarly, another compound of KOH is dissolved in de-ionized water and transforms to K^+ and OH^- ions (Equation (2)). Ti metals were dissolved and lost four electrons through MAO. Thereby, it transformed to positively charged cationic Ti^{4+} ions (Equation (3)). Synchronically, O_2 gaseous is released. The O_2 will either evolve as a gas, or dissolve into the solution as atoms and ionize to O^{2-}. Also, positively charged Ti^{4+} and negatively charged O^{2-} and/or OH^- ions react with each other due to the electrostatic interaction through MAO (Equations (4) and (5)). Eventually, the anodic oxidation reactions occur between cationic Ti^{4+} and anionic O^{2-}/OH^- ions on Ti substrate through the MAO. So, TiO_2 phase structure forms on cp-Ti substrate. The Gibbs energy of the anatase/rutile transition is negative because the phase of anatase is thermodynamically unstable for all temperature values. Anatase/rutile transition begins above 880 K and a completes at 1190 K [59]. The anatase to rutile transformation facilitates because the local temperature in micro discharge channels can reach 8000 K due to electron collisions through the MAO process [60]. Therefore, the amount of rutile increases with increasing MAO parameters such as treatment time, voltage and current.

Dissolution reactions in electrolyte:

$$Na_3PO_4 \rightarrow Na^+ + PO_4^{3-} \tag{1}$$

$$KOH \rightarrow K^+ + OH^- \tag{2}$$

$$Ti \rightarrow Ti^{4+} + 4e^- \tag{3}$$

Anodic oxidation reactions through MAO process:

$$Ti^{4+} + 4OH^- \rightarrow TiO_2 + 2H_2O \tag{4}$$

$$Ti^{4+} + 2O^{2-} \rightarrow TiO_2 \tag{5}$$

3.2. Surface Structures of the Coatings

The surface morphologies of the MAO and Cu-based MAO coatings on cp-Ti were evaluated by SEM as shown in Figure 3. As seen in Figure 3, the surfaces of both coatings are porous and rough. The MAO coatings contain many crater-like or volcano-like micro pores and a few micro cracks due to the existence of thermal stresses during whole process. The various-sized micro pores on the MAO surface are called as micro discharge channels occur by micro spark discharge. The micro sparks form at weak regions such as sites and edges on cp-Ti substrate during oxide film due to the existence of dielectric breakdown at the initial steps of the MAO process. Thus, these resulted in the increase of the intensity of micro discharge channels throughout the MAO process. The micro discharge channels have various sizes and occur on the substrate materials though the MAO process. So, volcano-like structures are observed on the MAO surfaces. Molten oxide structures term as oxide magma occur in micro discharge channels owing to the existence of local high temperature (up to about 10^4 K) and high pressure (approximately 100 MPa) [61,62]. The molten oxides in micro discharge channels is rapidly cooled and solidified because it comes into contact with the electrolyte during MAO. Eventually, it stacked instantaneously to form the MAO coatings. These porous and rough surfaces are beneficial to cell attachment and lead to increased cell adhesion [63]. Moreover, it is reported that porous TiO_2 layers that promote the sinking of liquid into the pores owing to capillary forces are favorable for the seeding and spreading of cells [64,65].

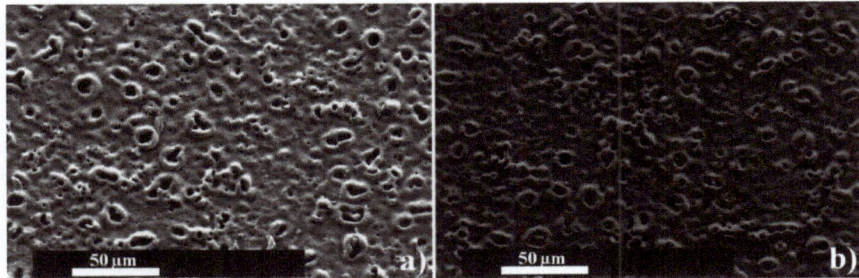

Figure 3. Surface SEM morphologies of the coatings: (**a**) the MAO coating and (**b**) Cu-based MAO coating.

The morphologies of both coating surfaces are nearly identical as shown in Figure 3, although TE treatment is applied on the MAO surface. The Cu layer has approximately 5 nm thickness and could not change the MAO surface or fill porous structure. In our previous studies [66,67], the hydroxyapatite-based MAO surface on zirconium was coated by anti-bacterial silver and zinc elements with 20 nm thickness and the surface morphology could not be changed. However, the surface chemistry, hydrophilicity/hydrophobicity, apatite forming ability and antibacterial activity are changed although morphologies of the MAO coating surfaces are maintained after the TE process. The 3-D mapping average roughness values of the MAO and Cu-based MAO surfaces are obtained as 1.10 and 1.16 µm, respectively. Thus, the average roughness of both surfaces was not significantly changed.

3.3. Elemental Chemical Analysis of the Coatings

The elemental analysis spectra results (the elemental amounts) on the MAO and Cu-based MAO coatings were analyzed by EDS and are shown in Table 2. A trace of Cu element was detected on the Cu-based surface while the elements of Ti, O and P were observed on both surfaces. The Ti peaks originate from the substrate material and TiO_2 structure in MAO coating. Furthermore, the O and P elements consist of anionic compounds such as PO_4^{3-} and OH^- in Na_3PO_4 and KOH-based alkaline MAO electrolyte. Anionic compounds migrate from electrolyte to substrate due to the existence of opposite charged ions under electrical field through MAO. Afterwards, they react with positively charged cationic ions (Ti^{4+}) accumulate on the MAO coating. However, the P element is not a crystalline form as seen in Figures 1 and 2 because it could not be detected by powder- or TF-XRD. The existence of the possible Ti-P-based compounds could not be proven in Figures 1 and 2. It can be expressed that they could not be transformed from amorphous to crystalline form on the surfaces whereas Ti^{4-} and PO_4^{3-} react with each other during MAO as supported by Figures 1 and 2. So, it could be stated that P-based compounds are an amorphous structure on the MAO-based surfaces. Moreover, the Cu was homogeneously stratified at a nanometer scale through the whole surface by TE process in that it is deposited on the MAO surface. The existence of the Cu element on the MAO surface is verified by the EDS-area whereas the crystallinity of it could not be confirmed by TF-XRD as shown in Figure 2.

Thus, it is observed as a trace amount on the MAO and TE combined surface. The Cu alters the surface chemistry of the MAO without any morphological changing as shown in Figure 3. Furthermore, it can be concluded that the Cu film on the MAO surface is an amorphous structure.

Table 2. EDS area spectra results of the MAO and the Cu-based MAO coatings.

Elements	The MAO Coating		The Cu-Based MAO Coating	
	wt.%	at.%	wt.%	at.%
O	34.82	60.03	33.06	58.23
P	7.76	6.91	8.53	7.76
Ti	57.42	33.06	55.94	32.91
Cu	–	–	2.48	1.10

3.4. Wettability of the Coatings

The wettability (hydrophilicity/hydrophobicity) of both surfaces was evaluated by a CAG device as shown in Table 3. Also, the average contact angle values of the coatings dependent on contacting time were given in Table 3 after the water droplet made contact with the surface. The CAG measurement is an efficient method to stay on top of the surface wettability and the surface free energy [68]. A small contact angle value refers to good wettability as a high contact angle value indicates a poor wettability. For orthopedic and dental implant applications, a good surface hydrophilicity is necessary for adherent growth of cell and tissue and it represents a good biocompatibility [68–71]. Furthermore, this was supported by in vitro apatite-forming ability results (in Section 3.5).

Table 3. The average contact angle values of the MAO and Cu-based MAO surfaces at post-contacting time of droplet.

The Droplet' Contacting Time (s)	The Contact Angle Values of the Substrates (°)	
	MAO Coating	Cu-Based MAO Coating
0	93.9	94.1
10	86.8	85.3
20	85.2	82.9
30	84.3	81.5
40	83.8	80.4
50	83.1	79.8
60	82.9	79.2
70	82.4	78.8
80	81.9	78.2
90	81.4	77.6

The surface wettability depends on many factors such as surface morphology, surface chemistry, roughness etc. [72,73]. The volcano-like pores on the surface absorb contacted distilled water by owing to capillary forces. So, for both surfaces, the contact angle values gradually decrease with increasing contact time as expected. Also, both surfaces exhibit hydrophilic characteristics because the contact angle values are lower than 90° [74]. It is clearly stated that TiO_2-based MAO surfaces are rough, indicating hydrophilic properties [75].

There is no extreme difference in contact angle values of both coatings, as shown in Table 3. However, it is obvious that the Cu-based surface is more hydrophilic than the MAO surface. In this study, the wettability of the surfaces change, whereas the surface morphologies of them are nearly identical. In a previous study [66], the surface of Zn-based with 5 nm thickness hydroxyapatite-based coating on zirconium was super hydrophilic with respect to the plain MAO surface. Similarly, another study [67], the surface of Ag-based with 20 nm thickness hydroxyapatite-based coating on zirconium was observed more hydrophobic than the one of plain MAO surface. So, it could be concluded that the hydrophilic/hydrophobic nature of the surfaces are strongly related to the surface chemistry even if the surfaces have identical morphology. Moreover, it is clear that the hydrophilicity of the MAO surface is improved after the Cu with 5 nm thickness is deposited on the MAO surface. Essentially, these two different surfaces' wettability is related to the polarity of them. The polar surfaces indicate that hydrophilicity/lower contact angles improve the wettability, while the opposite trend is observed in non-polar surfaces [76]. Eventually, it could be concluded that the accumulation of Cu onto the MAO increases the wettability/hydrophilicity. Also, this situation is beneficial for the attachment of cell and tissue for medical applications.

3.5. In Vitro Bioactivity of the Coatings

It is claimed that newly formed bone-like apatite layers on its surface under living body conditions is an essential requirement for binding bone tissue of the implant materials. This situation refers to in vivo apatite formation on the implant materials. An apatite structure on the surface can occur by immersion in SBF up to week-long periods under body temperature (36.5/37 °C) except for in vivo experimental conditions. The apatite formation/apatite forming ability on the surface provides predictive information about in vitro bioactivity. Thus, in vitro apatite forming ability on the surface is an important assessment to evaluate the bioactivity of the implants. However, "Apatite-forming ability is just a necessary but by no means sufficient precondition of "bioactivity". "Bioactivity" is a very complex interplay of many factors, where apatite-forming ability is just one of many" [77]. So, in order to be predicting information about bioactivity of all coating surfaces, in vitro immersion test was carried out under SBF condition at 36.5 °C for 28 days. After immersion tests were completed, the morphologies, elemental amount, phase structure and functional groups of all surfaces were characterized by SEM, EDS-area, TF-XRD and FTIR, respectively as seen in Figure 4, Table 4, Figures 5 and 6.

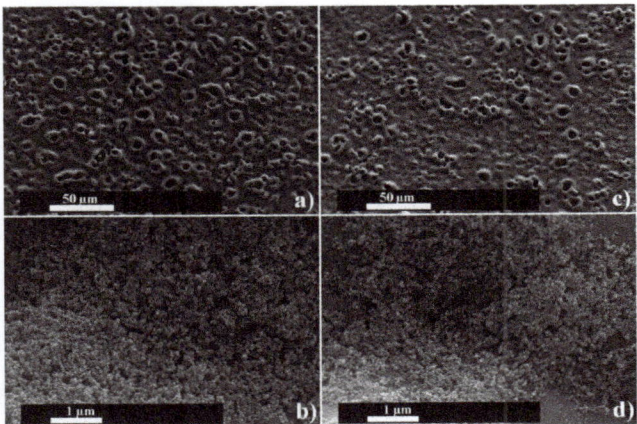

Figure 4. Surface morphologies of the coatings at post-immersion in SBF at 36.5 °C for 28 days: (**a**) 500× and (**b**) 20000× for the MAO surface and (**c**) 500× and (**d**) 20000× for the Cu-based MAO surface.

Table 4. EDS area spectra results of the MAO and the Cu-based MAO coatings at post-immersion in SBF.

Elements	The MAO Coating		The Cu-Based MAO Coating	
	wt.%	at.%	wt.%	at.%
O	43.82	68.48	37.88	63.15
P	7.28	5.88	7.27	6.26
Ca	1.27	0.79	0.85	0.56
Ti	47.63	24.86	53.73	29.91
Cu	–	–	0.27	0.11

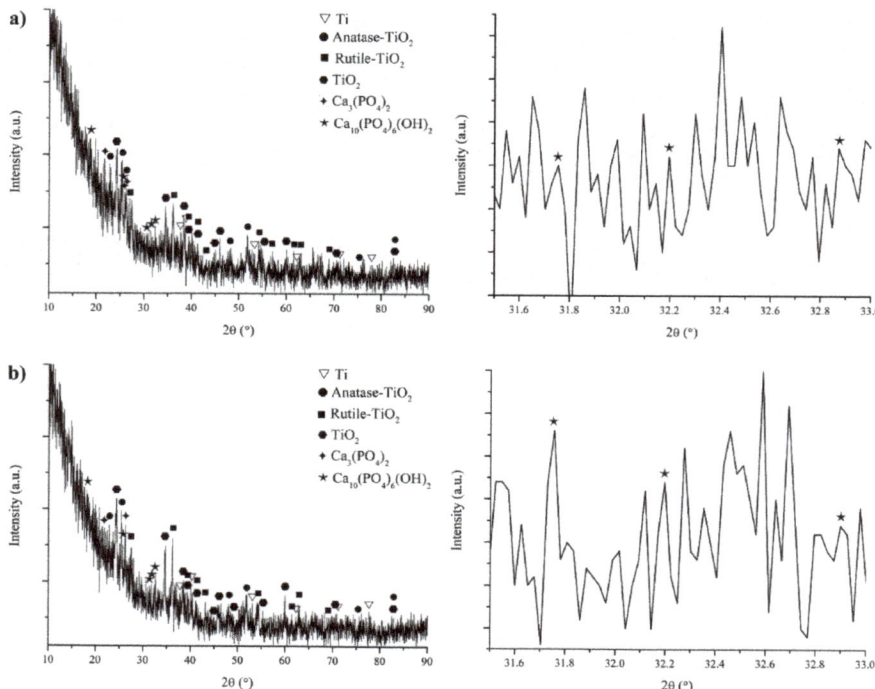

Figure 5. TF-XRD spectra pattern of the coatings at post-immersion in SBF at 36.5 °C for 28 days: (**a**) the MAO coating and (**b**) Cu-based MAO coating.

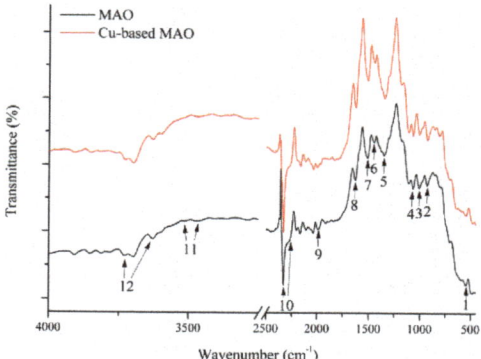

Figure 6. FTIR spectra of the coatings at post-immersion in SBF at 36.5 °C for 28 days: (black line) the MAO coating and (red line) Cu-based MAO coating.

After immersion in SBF for 28 days, the morphologies of both coating surfaces were investigated by SEM as shown in Figure 4. It could be clearly observed that the fine-dispersed particles are well dispersed throughout the whole surface at the post-immersion process. These fine-dispersed particles are nearly coated on the surfaces as seen in Figure 4. These particles are well dispersed through the whole surface. However, it is observed that the dispersed particles formed on the Cu-based surface are a little thinner than the one on the MAO surface at micron scales. It is well known that the crystalline apatite structure formed on the surfaces refers to predicting bioactivity. Therefore, as seen in TF-XRD and FTIR spectra results, the existence of crystalline apatite structure on the Cu-based MAO surface are greater than one the MAO surface whereas the apatite layer on the Cu-based MAO surface is a little thin respect to the MAO surface at macron scales scanning ($10^6 \times 10^6$ µm^2). So, it can be stated that the Cu-nano layer on the MAO surface triggers increasing the crystallinity of apatite structure under SBF.

In order to get information, the elemental structures and the elemental amounts of both surfaces at post-immersion in SBF, the MAO and Cu-based MAO surfaces were analyzed by the EDS-area as shown in Table 4. In addition to the existence of Ti, O, P and Cu detected on the surfaces at pre-immersion in SBF as given in Table 2, an extra Ca element is obtained on both surfaces at post-immersion in SBF as given in Figure 5 and Table 4. The presence of this is connected with the diffusion Ca^{2+} ions of Ca-based SBF surfaces onto the surfaces during the immersion process. Positively charged Ca^{2+} ions in SBF migrate on the Ti-OH based surfaces due to the electrostatic interactions of oppositely charged ions. Then, PO$_4^{3-}$ and OH$^-$ ions in SBF diffuse to both surfaces. These migrations and reactions are carried out through the immersion process in SBF. Eventually, they formed an apatite layer on the surfaces during the post-immersion process. Thus, an extra Ca element is observed on both surfaces during the post-immersion process. However, the electron penetration depth varies from 0.2 to 2 µm, depending on the accelerating voltage in EDS analysis. Thus, only presented elements in these particles formed at the post-immersion process in SBF cannot be observed by EDS. Besides the elements in these particles, the elements such as Ti, O, P and Cu existed in the MAO and PVD layer were observed by EDS analysis.

The phase structures of these newly formed dispersed structures were characterized by TF-XRD and FTIR as shown in Figures 5 and 6. The phases of Ti (JCPDS card number: 04-004-8480), anatase (JCPDS card number: 01-083-5914), rutile (JCPDS card number: 04-006-8034), TiO$_2$ (JCPDS card number: 01-070-2556), TCP (tri-calcium phosphate: JCPDS card number: 044-1294)) and apatite (JCPDS card number: 00-009-0432) on the MAO and Cu-based MAO surfaces at post-immersion in SBF were observed in Figure 5a,b, respectively. The existences of Ti, TiO$_2$, anatase and rutile on both surfaces have been verified in Figures 1 and 2 at pre-immersion in SBF. The phases of TCP and apatite on both surfaces form at post-immersion in SBF under body temperature for 28 days. So, the fine-dispersed particles monitored in Figure 4 refer to TCP and apatite phases on immersed surfaces. As seen in Figure 5a,b, the intensity and amount of crystalline apatite structure on the Cu-based MAO are greater than ones on the MAO whereas the elemental amounts of Ca on the Cu-based MAO are lower than one on the MAO as reported in EDS-area spectra and amount results. The apatite-forming ability on the Cu-based MAO surface is high compared to the MAO surface according to TF-XRD spectra.

The functional groups and phases on the MAO and Cu-based MAO coatings at post-immersion in SBF were investigated by FTIR as seen in Figure 6. Also, at post-immersion in SBF, FTIR spectrum analysis results such as wave numbers, band modes, band assignments and phase were presented in Table 5. The band structures obtained on both surfaces at post-immersion in SBF are (PO$_4^{3-}$), (C–C), (CO$_3^{2-}$), (OH$^-$), (P–H) and (Ti–O–Ti). The FTIR spectra curves found at 560 and 1026, 962 and 1050, 1403, 1420–1425 and 1460–1465, 1641, 1980 and 2302–2388, 3397–3448 and 3640–3742 cm^{-1} correspond to (PO$_4^{3-}$), (PO$_4^{3-}$), (C–C$^-$), (CO$_3^{2-}$), (OH$^-$), (P–H), (Ti–O–Ti) and (OH$^-$), respectively [22,43,78–95]. The ATR-FTIR peaks of both surfaces verify the presence of (C–C) at 1403 cm^{-1} [43,78–80], the presence of (OH$^-$) ion at 1641 cm^{-1} [43,78,79] and the presence of (Ti–O–Ti) in the regions of 3397–3448 cm^{-1} [43,78–80]. In addition to XRD spectra, the presence of TiO$_2$ structure on both surfaces are indicated by the bands of (C–C), (OH$^-$) and (Ti–O–Ti) on FTIR once again [43,78–80]. The FTIR spectra curves on both surfaces refer to the presence of (PO$_4^{3-}$) at 560 cm^{-1} [81–83] and 1026 cm^{-1} [83–86], the presence of (PO$_4^{3-}$) at 962 cm^{-1} [83–87] and 1050 cm^{-1} [83–87], the presence of (P–H) in the regions of 1980 cm^{-1} [88–93] and 2302–2388 cm^{-1} [88–93] and the presence of (OH$^-$) at 3640–3742 cm^{-1} [88,91–94]. The apatite and TCP structures formed on both surfaces at post-immersion in SBF are proved by the presence of (PO$_4^{3-}$), (P–H) and (OH$^-$) bands on FTIR curves [81–94]. The FTIR spectra curves on both surfaces verify the presence of (CO$_3^{2-}$) at 1420–1425 cm^{-1} [22,95] and 1460–1465 cm^{-1} [22,95]. The substituted carbonated-apatite formed on both surfaces at post-immersion in SBF is proved by the presence of (CO_3^{2-}) [22,95]. It is clear that the induced apatite is a carbonated apatite at post-immersion in SBF for 28 days. It is well known that the sharp and deep of FTIR peaks provide information about the crystallinity of phases on the surface [96]. As supported in Figure 5, the sharp and deep peaks of (PO_4^{3-}) bands in the Cu-based MAO verify the existence of highly crystalline apatite structure respect to the MAO in Figure 6. Thus, it can be stated that the Cu on the MAO improve predicting bioactivity compared to the MAO surface.

Table 5. ATR-FTIR spectrum analysis results for the MAO and Cu-based MAO coatings after immersion in SBF.

Peak Number	Wavenumber (cm^{-1})	Band Assignment	Band Mode	Phase Structures	References
1	560	PO_4^{3-}	Stretching	Apatite	[81–83]
2	962	PO_4^{3-}	Stretching	Apatite, TCP	[83–87]
3	1026	PO_4^{3-}	Stretching	Apatite	[83–86]
4	1050	PO_4^{3-}	Stretching	Apatite, TCP	[83–87]
5	1403	C–C	Stretching	TiO_2	[43,78–80]
6	1420–1425	CO_3^{2-}	Stretching	A-type apatite	[22,95]
7	1460–1465	CO_3^{2-}	Stretching	B-type apatite	[22,95]
8	1641	OH^-	Stretching	TiO_2	[43,78,79]
9	1980	P–H	Stretching	Apatite	[88–93]
10	2302–2388	P–H	Stretching	Apatite	[88–93]
11	3397–3448	Ti–O–Ti	Stretching	TiO_2	[43,78–80]
12	3640–3742	OH^-	Stretching	Apatite	[88,91–94]

SBF is a metastable calcium phosphate-based electrolyte supersaturated compared to the apatite structure [97]. However, it is stated that a chemical stimulus is required to trigger the heterogeneous nucleation of apatite from the SBF because the homogeneous nucleation threshold of apatite is very high [98]. The hydroxyl groups such as Ti–OH on the surfaces are basically essential to induce the apatite nucleation. The provision of abundant Ti–OH groups and the enrichment of calcium and phosphate trigger the nucleation of apatite on the MAO surface [97]. After apatite nuclei forms on the surface, the ions of Ca^{2+}, PO_4^{3-} and CO_3^{2-} in SBF diffuse to TiO_2-based surfaces to combine with apatite nuclei owing to the electrostatic interactions of opposite charged ions. As a result, a novel apatite layer is formed on both coating surfaces.

3.6. In Vitro Antibacterial Activity of the Coatings

In order to determine the antibacterial contribution of the Cu layer on the MAO process, the level of bacterial colony adhering to the cp-Ti, the MAO (TiO_2) and Cu-based MAO surfaces was investigated. Figure 7 shows active colony ratios of gram-positive (S. aureus) and gram-negative (E. coli) bacteria adhered to all tested surfaces. Also, Figures 8 and 9 show S. aureus and E. coli colony plates formed by bacteria adhering to all surfaces after incubation, respectively. It was determined that the adhesion of gram-positive and gram-negative bacteria on the cp-Ti surfaces was lower than that of the MAO surface. The MAO surfaces containing bioactive and biocompatible TiO_2 structures are porous and rough. Thus, the surface energy of the MAO is greater than one of the smooth surfaces. Hence, it can be concluded that the active colony ratios of S. aureus and E. coli adhered on the MAO surfaces increased in very small amount compared to the smooth cp-Ti surfaces. However, this difference was not statistically significant ($p > 0.05$). So, the numbers of bacteria adhering to cp-Ti and the MAO surfaces were very close to each other.

The antibacterial activity of the Cu-based MAO surfaces against E. coli was determined as 68.0%. Also, the antibacterial activity of the Cu-based MAO surfaces against S. aureus was observed as 69.6%, respectively. It has been observed that Cu-based MAO coating process provides antibacterial property to the substrate and MAO surfaces. Also, the Cu coating increases the antibacterial property by 1.11-1.18-fold against S. aureus and E. coli, respectively. This result shows that the Cu-based MAO coating on the surface has a significant effect on antibacterial activity. The increase in antibacterial activity by increasing the coating can be explained by the more intense interaction of Cu ions on the surface and bacteria.

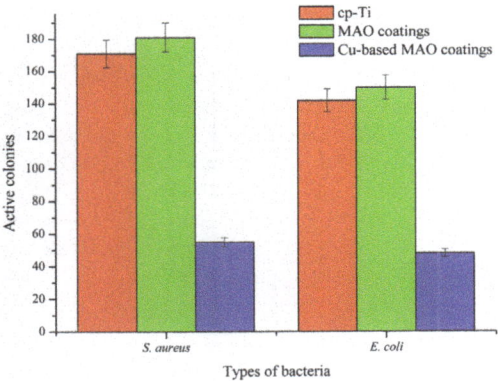

Figure 7. Active colony ratios of *S. aureus* as gram positive bacteria and *E. coli* as gram negative bacteria on cp-Ti substrate, the MAO and Cu-based MAO surface cultivation.

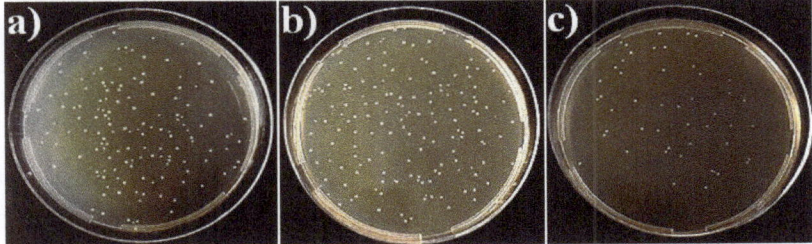

Figure 8. Culture plate photographs of *S. aureus* after re-cultivation: (**a**) cp-Ti substrate, (**b**) the MAO and (**c**) Cu-based MAO surface cultivation.

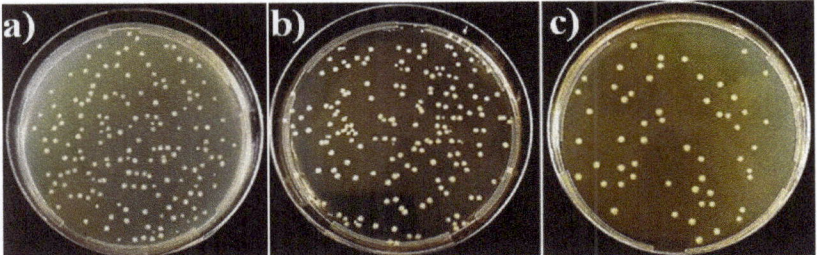

Figure 9. Culture plate photographs of *E. coli* after re-cultivation: (**a**) cp-Ti substrate, (**b**) the MAO and (**c**) Cu-based MAO surface cultivation.

The increase in antibacterial activity of surfaces after Cu-based coating may be explained by the toxic effect of Cu ions on the bacterial cell. In mediums containing Cu-based surfaces, bacteria are immobilized to the coated surfaces; proliferation and movement are restricted, and cell death occurs [99]. Copper causes toxic effects on the bacteria by multiple mechanisms and prevents the resistance formation in the bacteria. The first mechanism is that copper ions deform the cell wall and cause cell death. In another mechanism, copper ions inactivate membrane proteins and enzymes, disrupt transporter molecules and lead to cell death [100,101]. Previously, some researchers have investigated the antibacterial properties of Cu-based MAO surfaces against various microorganisms [102,103]. Trapalis et al. [104] reported that copper-coated surfaces (Cu/SiO$_2$) exhibited significant antibacterial activity against *E. coli*.

Another important observation in this study is that Cu-based surfaces show higher antibacterial activity against *S. aureus* than *E. coli*. The Cu-based surfaces, which exhibit the highest antibacterial activity in this study, showed 1.02 times higher antibacterial activity against *S. aureus* compared to *E. coli*. This result can be explained by the structural differences of gram-positive and gram-negative cells. Gram-negative bacteria have an outer membrane in addition to the cell wall, while gram-positive bacteria do not have an outer membrane. The outer membrane acts as a barrier to reduce the transport of metals to the bacteria. For this reason, copper penetrates rapidly into the cell in gram-positive cells and a serious toxic effect occurs. However, in gram-negative bacteria containing outer membrane, copper transport to the cell is restricted; the formation of toxic effect is delayed and reduced [50,105]. Similar studies in the literature support these results. Wang et al. [105] reported that selenium-coated surfaces are more effective against gram-positive bacteria and that this result is due to the cellular difference between *S. aureus* and *E. coli*.

4. Conclusions

A novel Cu-based TiO$_2$ coating has been produced on cp-Ti surfaces by using combined two-step MAO and PVD-TE methods for dental implant applications. The TF-XRD spectra results showed that the existence of crystalline Cu on the MAO surfaces could not be observed whereas it was detected as elemental structure by EDS attached to SEM. Also, the SEM analyses indicated that the surface morphologies of the MAO and Cu-based MAO coatings were porous and rough. The morphology and topography of both coating surfaces were not changed by PVD-TE treatment. However, the hydrophilicity of Cu-based MAO surfaces was improved with respect to the MAO surfaces owing to the enhancing polarity of Cu on the MAO surface. Moreover, in vitro bioactivity of Cu-based TiO$_2$ surfaces was increased compared to the MAO surfaces after immersion test in SBF for 28 days. These were supported by TF-XRD, ATR-FTIR and SEM as written in Section 3.5. Furthermore, it was observed that the amounts of active-bacteria colonies lived on Cu-based MAO surfaces were lower than ones of the MAO surface for gram positive (*S. aureus*) and gram negative (*E. coli*).

There have been some remarkable studies on the production of antibacterial Cu/CuO-nanoparticles containing Cu-incorporated- and Cu-containing-TiO$_2$ coatings on Ti by a single or hybrid MAO methods [42,45,51–56]. However, Cu nanoparticles were randomly separated through the whole MAO surface in aforementioned studies. Also, Cu was not uniformly separated on Cu-incorporated- and Cu-containing-MAO surfaces. So, antibacterial adhesion properties on the Cu-MAO surface could not be sufficiently improved. Moreover, the Cu was observed around micro discharge channels and could not be dispersed during combined hybrid MAO surfaces. Moreover, in vitro bioactivity of the Cu-nanoparticles, Cu-incorporated and Cu-containing-MAO coatings were not investigated although in vitro biocompatibility investigations were carried out in aforementioned studies. In conclusion, it can be stated that the Cu-based TiO$_2$ coatings are porous, rough and have the potential for biomedical surface coating applications due to their hopeful properties such as surface chemistry, morphology, hydrophilicity, in vitro bioactivity and in vitro antibacterial resistance with respect to other literature studies.

Funding: This research received no external funding.

Acknowledgments: The author would like to special thank F. Unal for running Physical Vapor Deposition Thermal Evaporation System at Giresun University GRUMLAB, A. Nazim for running SEM and EDS, A. Sen for running XRD at Gebze Technical University and B. Alcan and Y. Ozturk for running TF-XRD at TUBITAK MAM Materials Institute.

Conflicts of Interest: The author declares no conflict of interest.

References

1. Quintero, D.; Galvis, O.; Calderón, J.A.; Castaño, J.G.; Echeverría, F. Effect of electrochemical parameters on the formation of anodic films on commercially pure titanium by plasma electrolytic oxidation. *Surf. Coat. Technol.* **2014**, *258*, 1223–1231. [CrossRef]
2. Yang, W.; Xu, D.; Guo, Q.; Chen, T.; Chen, J. Influence of electrolyte composition on microstructure and properties of coatings formed on pure Ti substrate by micro arc oxidation. *Surf. Coat. Technol.* **2018**, *349*, 522–528. [CrossRef]
3. Teker, D.; Muhaffel, F.; Menekse, M.; Karaguler, N.G.; Baydogan, M.; Cimenoglu, H. Characteristics of multi-layer coating formed on commercially pure titanium for biomedical applications. *Mater. Sci. Eng. C* **2015**, *48*, 579–585. [CrossRef] [PubMed]
4. Liu, J.-X.; Yang, D.-Z.; Shi, F.; Cai, Y.-J. Sol–gel deposited TiO_2 film on NiTi surgical alloy for biocompatibility improvement. *Thin Solid Films* **2003**, *429*, 225–230. [CrossRef]
5. Duarte, L.T.; Bolfarini, C.; Biaggio, S.R.; Rocha-Filho, R.C.; Nascente, P.A.P. Growth of aluminum-free porous oxide layers on titanium and its alloys Ti-6Al-4V and Ti-6Al-7Nb by micro-arc oxidation. *Mater. Sci. Eng. C* **2014**, *41*, 343–348. [CrossRef] [PubMed]
6. Terleeva, O.P.; Sharkeev, Y.P.; Slonova, A.I.; Mironov, I.V.; Legostaeva, E.V.; Khlusov, I.A.; Matykina, E.; Skeldon, P.; Thompson, G.E. Effect of microplasma modes and electrolyte composition on micro-arc oxidation coatings on titanium for medical applications. *Surf. Coat. Technol.* **2010**, *205*, 1723–1729. [CrossRef]
7. Yoon, I.-K.; Hwang, J.-Y.; Jang, W.-C.; Kim, H.-W.; Shin, U.S. Natural bone-like biomimetic surface modification of titanium. *Appl. Surf. Sci.* **2014**, *301*, 401–409. [CrossRef]
8. Albrektsson, T.; Jacobsson, M. Bone-metal interface in osseointegration. *J. Prosthet. Dent.* **1987**, *57*, 597–607. [CrossRef]
9. Venkateswarlu, K.; Rameshbabu, N.; Sreekanth, D.; Bose, A.C.; Muthupandi, V.; Subramanian, S. Fabrication and characterization of micro-arc oxidized fluoride containing titania films on cp Ti. *Ceram. Int.* **2013**, *39*, 801–812. [CrossRef]
10. Durdu, S.; Usta, M.; Berkem, A.S. Bioactive coatings on Ti6Al4V alloy formed by plasma electrolytic oxidation. *Surf. Coat. Technol.* **2016**, *301*, 85–93. [CrossRef]
11. Drnovšek, N.; Rade, K.; Milačič, R.; Štrancar, J.; Novak, S. The properties of bioactive TiO_2 coatings on Ti-based implants. *Surf. Coat. Technol.* **2012**, *209*, 177–183. [CrossRef]
12. He, X.; Zhang, G.; Wang, X.; Hang, R.; Huang, X.; Qin, L.; Tang, B.; Zhang, X. Biocompatibility, corrosion resistance and antibacterial activity of TiO_2/CuO coating on titanium. *Ceram. Int.* **2017**, *43*, 16185–16195. [CrossRef]
13. Çomaklı, O.; Yazıcı, M.; Kovacı, H.; Yetim, T.; Yetim, A.F.; Çelik, A. Tribological and electrochemical properties of TiO_2 films produced on cp-Ti by sol-gel and silar in bio-simulated environment. *Surf. Coat. Technol.* **2018**, *352*, 513–521. [CrossRef]
14. Uhm, S.-H.; Song, D.-H.; Kwon, J.-S.; Im, S.-Y.; Han, J.-G.; Kim, K.-N. Time-dependent growth of TiO_2 nanotubes from a magnetron sputtered Ti thin film. *Thin Solid Films* **2013**, *547*, 181–187. [CrossRef]
15. Garcia-Valenzuela, A.; Alvarez, R.; Rico, V.; Cotrino, J.; Gonzalez-Elipe, A.R.; Palmero, A. Growth of nanocolumnar porous TiO_2 thin films by magnetron sputtering using particle collimators. *Surf. Coat. Technol.* **2018**, *343*, 172–177. [CrossRef]
16. Cabanas-Polo, S.; Boccaccini, A.R. Electrophoretic deposition of nanoscale TiO_2: Technology and applications. *J. Eur. Ceram. Soc.* **2016**, *36*, 265–283. [CrossRef]
17. An, S.-H.; Matsumoto, T.; Miyajima, H.; Sasaki, J.-I.; Narayanan, R.; Kim, K.-H. Surface characterization of alkali- and heat-treated Ti with or without prior acid etching. *Appl. Surf. Sci.* **2012**, *258*, 4377–4382. [CrossRef]
18. Kim, S.-Y.; Kim, Y.-K.; Jang, Y.-S.; Park, I.-S.; Lee, S.-J.; Jeon, J.-G.; Lee, M.-H. Bioactive effect of alkali-heat treated TiO_2 nanotubes by water or acid treatment. *Surf. Coat. Technol.* **2016**, *303*, 256–267. [CrossRef]
19. Weng, F.; Chen, C.; Yu, H. Research status of laser cladding on titanium and its alloys: A review. *Mater. Des.* **2014**, *58*, 412–425. [CrossRef]
20. Sharifi, N.; Pugh, M.; Moreau, C.; Dolatabadi, A. Developing hydrophobic and superhydrophobic TiO_2 coatings by plasma spraying. *Surf. Coat. Technol.* **2016**, *289*, 29–36. [CrossRef]

21. Wang, H.-D.; He, P.-F.; Ma, G.-Z.; Xu, B.-S.; Xing, Z.-G.; Chen, S.-Y.; Liu, Z.; Wang, Y.-W. Tribological behavior of plasma sprayed carbon nanotubes reinforced TiO_2 coatings. *J. Eur. Ceram. Soc.* **2018**, *38*, 3660–3672. [CrossRef]
22. Zhou, R.; Wei, D.; Cheng, S.; Li, B.; Wang, Y.; Jia, D.; Zhou, Y.; Guo, H. The structure and in vitro apatite formation ability of porous titanium covered bioactive microarc oxidized TiO_2-based coatings containing Si, Na and Ca. *Ceram. Int.* **2014**, *40*, 501–509. [CrossRef]
23. Durdu, S.; Usta, M. The tribological properties of bioceramic coatings produced on Ti6Al4V alloy by plasma electrolytic oxidation. *Ceram. Int.* **2014**, *40*, 3627–3635. [CrossRef]
24. Du, Q.; Wei, D.; Wang, Y.; Cheng, S.; Liu, S.; Zhou, Y.; Jia, D. The effect of applied voltages on the structure, apatite-inducing ability and antibacterial ability of micro arc oxidation coating formed on titanium surface. *Bioact. Mater.* **2018**, *3*, 426–433. [CrossRef] [PubMed]
25. He, Y.; Zhang, Y.; Shen, X.; Tao, B.; Liu, J.; Yuan, Z.; Cai, K. The fabrication and in vitro properties of antibacterial polydopamine-ll-37-POPC coatings on micro-arc oxidized titanium. *Colloids Surf. B Biointerfaces* **2018**, *170*, 54–63. [CrossRef] [PubMed]
26. Sun, L.; Berndt, C.C.; Gross, K.A.; Kucuk, A. Material fundamentals and clinical performance of plasma-sprayed hydroxyapatite coatings: A review. *J. Biomed. Mater. Res.* **2001**, *58*, 570–592. [CrossRef] [PubMed]
27. Manso, M.; Jiménez, C.; Morant, C.; Herrero, P.; Martínez-Duart, J.M. Electrodeposition of hydroxyapatite coatings in basic conditions. *Biomaterials* **2000**, *21*, 1755–1761. [CrossRef]
28. Lee, J.-H.; Kim, H.-E.; Shin, K.-H.; Koh, Y.-H. Electrodeposition of biodegradable sol-gel derived silica onto nanoporous TiO_2 surface formed on Ti substrate. *Mater. Lett.* **2011**, *65*, 1519–1521. [CrossRef]
29. Xu, L.; Wu, C.; Lei, X.; Zhang, K.; Liu, C.; Ding, J.; Shi, X. Effect of oxidation time on cytocompatibility of ultrafine-grained pure Ti in micro-arc oxidation treatment. *Surf. Coat. Technol.* **2018**, *342*, 12–22. [CrossRef]
30. Yilmaz, M.S.; Sahin, O. Applying high voltage cathodic pulse with various pulse durations on aluminium via micro-arc oxidation (MAO). *Surf. Coat. Technol.* **2018**, *347*, 278–285. [CrossRef]
31. Yang, W.; Xu, D.; Yao, X.; Wang, J.; Chen, J. Stable preparation and characterization of yellow micro arc oxidation coating on magnesium alloy. *J. Alloy. Compd.* **2018**, *745*, 609–616. [CrossRef]
32. Yang, W.; Xu, D.; Wang, J.; Yao, X.; Chen, J. Microstructure and corrosion resistance of micro arc oxidation plus electrostatic powder spraying composite coating on magnesium alloy. *Corros. Sci.* **2018**, *136*, 174–179. [CrossRef]
33. Zhang, D.; Ge, Y.; Liu, G.; Gao, F.; Li, P. Investigation of tribological properties of micro-arc oxidation ceramic coating on Mg alloy under dry sliding condition. *Ceram. Int.* **2018**, *44*, 16164–16172. [CrossRef]
34. Zhang, J.; Fan, Y.; Zhao, X.; Ma, R.; Du, A.; Cao, X. Influence of duty cycle on the growth behavior and wear resistance of micro-arc oxidation coatings on hot dip aluminized cast iron. *Surf. Coat. Technol.* **2018**, *337*, 141–149. [CrossRef]
35. Cui, L.-Y.; Liu, H.-P.; Zhang, W.-L.; Han, Z.-Z.; Deng, M.-X.; Zeng, R.-C.; Li, S.-Q.; Wang, Z.-L. Corrosion resistance of a superhydrophobic micro-arc oxidation coating on Mg-4Li-1Ca alloy. *J. Mater. Sci. Technol.* **2017**, *33*, 1263–1271. [CrossRef]
36. Fan, X.; Feng, B.; Di, Y.; Lu, X.; Duan, K.; Wang, J.; Weng, J. Preparation of bioactive TiO film on porous titanium by micro-arc oxidation. *Appl. Surf. Sci.* **2012**, *258*, 7584–7588. [CrossRef]
37. Zhang, C.L.; Zhang, F.; Song, L.; Zeng, R.C.; Li, S.Q.; Han, E.H. Corrosion resistance of a superhydrophobic surface on micro-arc oxidation coated Mg-Li-Ca alloy. *J. Alloy. Compd.* **2017**, *728*, 815–826. [CrossRef]
38. Wang, Y.; Yu, H.; Chen, C.; Zhao, Z. Review of the biocompatibility of micro-arc oxidation coated titanium alloys. *Mater. Des.* **2015**, *85*, 640–652. [CrossRef]
39. Wang, L.; Shi, L.; Chen, J.; Shi, Z.; Ren, L.; Wang, Y. Biocompatibility of Si-incorporated TiO_2 film prepared by micro-arc oxidation. *Mater. Lett.* **2014**, *116*, 35–38. [CrossRef]
40. Li, Y.; Wang, W.; Liu, H.; Lei, J.; Zhang, J.; Zhou, H.; Qi, M. Formation and in vitro/in vivo performance of "cortex-like" micro/nano-structured TiO_2 coatings on titanium by micro-arc oxidation. *Mater. Sci. Eng. C* **2018**, *87*, 90–103. [CrossRef]
41. Liu, X.Y.; Chu, P.K.; Ding, C.X. Surface modification of titanium, titanium alloys, and related materials for biomedical applications. *Mater. Sci. Eng. R-Rep.* **2004**, *47*, 49–121. [CrossRef]
42. Wu, Q.J.; Li, J.H.; Zhang, W.J.; Qian, H.X.; She, W.J.; Pan, H.Y.; Wen, J.; Zhang, X.L.; Liu, X.Y.; Jiang, X.Q. Antibacterial property, angiogenic and osteogenic activity of Cu-incorporated TiO_2 coating. *J. Mater. Chem. B* **2014**, *2*, 6738–6748. [CrossRef]

43. Durdu, S.; Aktug, S.L.; Korkmaz, K.; Yalcin, E.; Aktas, S. Fabrication, characterization and in vitro properties of silver-incorporated TiO$_2$ coatings on titanium by thermal evaporation and micro-arc oxidation. *Surf. Coat. Technol.* **2018**, *352*, 600–608. [CrossRef]
44. Hu, H.; Zhang, W.; Qiao, Y.; Jiang, X.; Liu, X.; Ding, C. Antibacterial activity and increased bone marrow stem cell functions of Zn-incorporated TiO$_2$ coatings on titanium. *Acta Biomater.* **2012**, *8*, 904–915. [CrossRef] [PubMed]
45. Yao, X.; Zhang, X.; Wu, H.; Tian, L.; Ma, Y.; Tang, B. Microstructure and antibacterial properties of Cu-doped TiO$_2$ coating on titanium by micro-arc oxidation. *Appl. Surf. Sci.* **2014**, *292*, 944–947. [CrossRef]
46. Wu, C.; Zhou, Y.; Xu, M.; Han, P.; Chen, L.; Chang, J.; Xiao, Y. Copper-containing mesoporous bioactive glass scaffolds with multifunctional properties of angiogenesis capacity, osteostimulation and antibacterial activity. *Biomaterials* **2013**, *34*, 422–433. [CrossRef] [PubMed]
47. Li, J.; Zhai, D.; Lv, F.; Yu, Q.; Ma, H.; Yin, J.; Yi, Z.; Liu, M.; Chang, J.; Wu, C. Preparation of copper-containing bioactive glass/eggshell membrane nanocomposites for improving angiogenesis, antibacterial activity and wound healing. *Acta Biomater.* **2016**, *36*, 254–266. [CrossRef]
48. Matsumoto, N.; Sato, K.; Yoshida, K.; Hashimoto, K.; Toda, Y. Preparation and characterization of β-tricalcium phosphate Co-doped with monovalent and divalent antibacterial metal ions. *Acta Biomater.* **2009**, *5*, 3157–3164. [CrossRef]
49. Zhang, X.; Huang, X.; Ma, Y.; Lin, N.; Fan, A.; Tang, B. Bactericidal behavior of Cu-containing stainless steel surfaces. *Appl. Surf. Sci.* **2012**, *258*, 10058–10063. [CrossRef]
50. Ruparelia, J.P.; Chatterjee, A.K.; Duttagupta, S.P.; Mukherji, S. Strain specificity in antimicrobial activity of silver and copper nanoparticles. *Acta Biomater.* **2008**, *4*, 707–716. [CrossRef]
51. Huang, Q.; Liu, X.; Zhang, R.; Yang, X.; Lan, C.; Feng, Q.; Liu, Y. The development of Cu-incorporated micro/nano-topographical bio-ceramic coatings for enhanced osteoblast response. *Appl. Surf. Sci.* **2019**, *465*, 575–583. [CrossRef]
52. Zhu, W.; Zhang, Z.X.; Gu, B.B.; Sun, J.Y.; Zhu, L.X. Biological activity and antibacterial property of nano-structured TiO$_2$ coating incorporated with Cu prepared by micro-arc oxidation. *J. Mater. Sci. Technol.* **2013**, *29*, 237–244. [CrossRef]
53. Zhang, X.; Li, J.; Wang, X.; Wang, Y.; Hang, R.; Huang, X.; Tang, B.; Chu, P.K. Effects of copper nanoparticles in porous TiO$_2$ coatings on bacterial resistance and cytocompatibility of osteoblasts and endothelial cells. *Mater. Sci. Eng. C* **2018**, *82*, 110–120. [CrossRef] [PubMed]
54. Huang, Q.; Li, X.; Elkhooly, T.A.; Liu, X.; Zhang, R.; Wu, H.; Feng, Q.; Liu, Y. The Cu-containing TiO$_2$ coatings with modulatory effects on macrophage polarization and bactericidal capacity prepared by micro-arc oxidation on titanium substrates. *Colloids Surf. B Biointerfaces* **2018**, *170*, 242–250. [CrossRef] [PubMed]
55. Zhang, L.; Guo, J.Q.; Huang, X.Y.; Zhang, Y.N.; Han, Y. The dual function of Cu-doped TiO$_2$ coatings on titanium for application in percutaneous implants. *J. Mater. Chem. B* **2016**, *4*, 3788–3800. [CrossRef]
56. Wu, H.B.; Zhang, X.Y.; Geng, Z.H.; Yin, Y.; Hang, R.Q.; Huang, X.B.; Yao, X.H.; Tang, B. Preparation, antibacterial effects and corrosion resistant of porous Cu-TiO$_2$ coatings. *Appl. Surf. Sci.* **2014**, *308*, 43–49. [CrossRef]
57. Kokubo, T.; Takadama, H. How useful is SBF in predicting in vivo bone bioactivity? *Biomaterials* **2006**, *27*, 2907–2915. [CrossRef] [PubMed]
58. Wang, Y.M.; Jiang, B.L.; Lei, T.Q.; Guo, L.X. Dependence of growth features of microarc oxidation coatings of titanium alloy on control modes of alternate pulse. *Mater. Lett.* **2004**, *58*, 1907–1911. [CrossRef]
59. Hanaor, D.A.H.; Sorrell, C.C. Review of the anatase to rutile phase transformation. *J. Mater. Sci.* **2011**, *46*, 855–874. [CrossRef]
60. Yerokhin, A.L.; Nie, X.; Leyland, A.; Matthews, A. Characterisation of oxide films produced by plasma electrolytic oxidation of a Ti–6Al–4V alloy. *Surf. Coat. Technol.* **2000**, *130*, 195–206. [CrossRef]
61. Li, Z.W.; Di, S.C. Microstructure and properties of MAO composite coatings containing nanorutile TiO$_2$ particles. *Surf. Rev. Lett.* **2017**, *24*, 1750115. [CrossRef]
62. Yerokhin, L.; Snizhko, L.O.; Gurevina, N.L.; Leyland, A.; Pilkington, A.; Matthews, A. Discharge characterization in plasma electrolytic oxidation of aluminium. *J. Phys. D-Appl. Phys.* **2003**, *36*, 2110–2120. [CrossRef]
63. Cimenoglu, H.; Gunyuz, M.; Kose, G.T.; Baydogan, M.; Ugurlu, F.; Sener, C. Micro-arc oxidation of Ti6Al4V and Ti6Al7Nb alloys for biomedical applications. *Mater. Charact.* **2011**, *62*, 304–311. [CrossRef]

64. Deng, F.L.; Zhang, W.Z.; Zhang, P.F.; Liu, C.H.; Ling, J.Q. Improvement in the morphology of micro-arc oxidised titanium surfaces: A new process to increase osteoblast response. *Mater. Sci. Eng. C-Mater. Boil. Appl.* **2010**, *30*, 141–147. [CrossRef]
65. Zhang, Z.X.; Sun, J.Y.; Hu, H.J.; Wang, Q.M.; Liu, X.Y. Osteoblast-like cell adhesion on porous silicon-incorporated TiO$_2$ coating prepared by micro-arc oxidation. *J. Biomed. Mater. Res. Part B-Appl. Biomater.* **2011**, *97B*, 224–234. [CrossRef] [PubMed]
66. Durdu, S.; Aktug, S.L.; Aktas, S.; Yalcin, E.; Usta, M. Fabrication and in vitro properties of zinc-based superhydrophilic bioceramic coatings on zirconium. *Surf. Coat. Technol.* **2018**, *344*, 467–478. [CrossRef]
67. Durdu, S.; Aktug, S.L.; Aktas, S.; Yalcin, E.; Cavusoglu, K.; Altinkok, A.; Usta, M. Characterization and in vitro properties of anti-bacterial Ag-based bioceramic coatings formed on zirconium by micro arc oxidation and thermal evaporation. *Surf. Coat. Technol.* **2017**, *331*, 107–115. [CrossRef]
68. Li, Y.D.; Wang, W.Q.; Duan, J.T.; Qi, M. A super-hydrophilic coating with a macro/micro/nano triple hierarchical structure on titanium by two-step micro-arc oxidation treatment for biomedical applications. *Surf. Coat. Technol.* **2017**, *311*, 1–9. [CrossRef]
69. Tang, W.X.; Yan, J.K.; Yang, G.; Gan, G.Y.; Du, J.H.; Zhang, J.M.; Liu, Y.C.; Shi, Z.; Yu, J.H. Effect of electrolytic solution concentrations on surface hydrophilicity of micro-arc oxidation ceramic film based on Ti6Al4V titanium alloy. *Rare Met. Mater. Eng.* **2014**, *43*, 2883–2888.
70. Zhang, Y.M.; Bataillon-Linez, P.; Huang, P.; Zhao, Y.M.; Han, Y.; Traisnel, M.; Xu, K.W.; Hildebrand, H.F. Surface analyses of micro-arc oxidized and hydrothermally treated titanium and effect on osteoblast behavior. *J. Biomed. Mater. Res. Part A* **2004**, *68A*, 383–391. [CrossRef] [PubMed]
71. Das, K.; Bose, S.; Bandyopadhyay, A. Surface modifications and cell–materials interactions with anodized Ti. *Acta Biomater.* **2007**, *3*, 573–585. [CrossRef] [PubMed]
72. Wang, Z.W.; Li, Q.; She, Z.X.; Chen, F.N.; Li, L.Q.; Zhang, X.X.; Zhang, P. Facile and fast fabrication of superhydrophobic surface on magnesium alloy. *Appl. Surf. Sci.* **2013**, *271*, 182–192. [CrossRef]
73. Wang, P.; Zhang, D.; Qiu, R.; Wan, Y.; Wu, J.J. Green approach to fabrication of a super-hydrophobic film on copper and the consequent corrosion resistance. *Corros. Sci.* **2014**, *80*, 366–373. [CrossRef]
74. Cui, X.J.; Lin, X.Z.; Liu, C.H.; Yang, R.S.; Zheng, X.W.; Gong, M. Fabrication and corrosion resistance of a hydrophobic micro-arc oxidation coating on AZ31 Mg alloy. *Corros. Sci.* **2015**, *90*, 402–412. [CrossRef]
75. Bayati, M.R.; Molaei, R.; Kajbafvala, A.; Zanganeh, S.; Zargar, H.R.; Janghorban, K. Investigation on hydrophilicity of micro-arc oxidized TiO$_2$ nano/micro-porous layers. *Electrochim. Acta* **2010**, *55*, 5786–5792. [CrossRef]
76. Pereira, M.M.; Kurnia, K.A.; Sousa, F.L.; Silva, N.J.O.; Lopes-da-Silva, J.A.; Coutinhoa, J.A.P.; Freire, M.G. Contact angles and wettability of ionic liquids on polar and non-polar surfaces. *Phys. Chem. Chem. Phys.* **2015**, *17*, 31653–31661. [CrossRef]
77. Kokubo, T. Bioactive glass ceramics: Properties and applications. *Biomaterials* **1991**, *12*, 155–163. [CrossRef]
78. Leon, A.; Reuquen, P.; Garin, C.; Segura, R.; Vargas, P.; Zapata, P.; Orihuela, P.A. FTIR and raman characterization of TiO$_2$ nanoparticles coated with polyethylene glycol as carrier for 2-methoxyestradiol. *Appl. Sci.* **2017**, *7*, 49. [CrossRef]
79. Chellappa, M.; Anjaneyulu, U.; Manivasagam, G.; Vijayalakshmi, U. Preparation and evaluation of the cytotoxic nature of TiO$_2$ nanoparticles by direct contact method. *Int. J. Nanomed.* **2015**, *10*, 31–41.
80. Khan, M.; Naqvi, A.H.; Ahmad, M. Comparative study of the cytotoxic and genotoxic potentials of zinc oxide and titanium dioxide nanoparticles. *Toxicol. Rep.* **2015**, *2*, 765–774. [CrossRef]
81. Ślósarczyk, A.; Paszkiewicz, Z.; Paluszkiewicz, C. FTIR and XRD evaluation of carbonated hydroxyapatite powders synthesized by wet methods. *J. Mol. Struct.* **2005**, *744–747*, 657–661.
82. Rapacz-Kmita, A.; Paluszkiewicz, C.; Ślósarczyk, A.; Paszkiewicz, Z. FTIR and XRD investigations on the thermal stability of hydroxyapatite during hot pressing and pressureless sintering processes. *J. Mol. Struct.* **2005**, *744–747*, 653–656. [CrossRef]
83. Nayak, Y.; Rana, R.; Pratihar, S.; Bhattacharyya, S. Low-temperature processing of dense hydroxyapatite–zirconia composites. *Int. J. Appl. Ceram. Technol.* **2008**, *5*, 29–36. [CrossRef]
84. Kim, H.-W.; Noh, Y.-J.; Koh, Y.-H.; Kim, H.-E.; Kim, H.-M. Effect of CaF$_2$ on densification and properties of hydroxyapatite–zirconia composites for biomedical applications. *Biomaterials* **2002**, *23*, 4113–4121. [CrossRef]

85. Basar, B.; Tezcaner, A.; Keskin, D.; Evis, Z. Improvements in microstructural, mechanical, and biocompatibility properties of nano-sized hydroxyapatites doped with yttrium and fluoride. *Ceram. Int.* **2010**, *36*, 1633–1643. [CrossRef]
86. Rintoul, L.; Byrne, E.W.; Suzuki, S.; Grondahl, L. FT–IR spectroscopy fluoro–substituted hydroxyapatite: Strengths and limitations. *J. Mater. Sci. Mater. Med.* **2007**, *18*, 1701–1709. [CrossRef]
87. Aykul, A. Investigation of Effect of Yttrium Fluoride on Microstructural and Mechanical Properties of Hydroxyapatite and Partially Stabilized Zirconia Composites. Master's Thesis, Gebze Institute of Technology, Gebze, Turkey, January 2010.
88. Aktuğ, S.L.; Durdu, S.; Yalçın, E.; Çavuşoğlu, K.; Usta, M. Bioactivity and biocompatibility of hydroxyapatite-based bioceramic coatings on zirconium by plasma electrolytic oxidation. *Mater. Sci. Eng. C* **2017**, *71*, 1020–1027. [CrossRef]
89. Aktuğ, S.L.; Durdu, S.; Yalçın, E.; Çavuşoğlu, K.; Usta, M. In vitro properties of bioceramic coatings produced on zirconium by plasma electrolytic oxidation. *Surf. Coat. Technol.* **2017**, *324*, 129–139. [CrossRef]
90. Durdu, S.; Deniz, Ö.F.; Kutbay, I.; Usta, M. Characterization and formation of hydroxyapatite on Ti6Al4V coated by plasma electrolytic oxidation. *J. Alloy. Compd.* **2013**, *551*, 422–429. [CrossRef]
91. Pecheva, E.V.; Pramatarova, L.D.; Maitz, M.F.; Pham, M.T.; Kondyuirin, A.V. Kinetics of hydroxyapatite deposition on solid substrates modified by sequential implantation of Ca and P ions—Part I. FTIR and raman spectroscopy study. *Appl. Surf. Sci.* **2004**, *235*, 176–181. [CrossRef]
92. Shaltout, A.A.; Allam, M.A.; Moharram, M.A. FTIR spectroscopic, thermal and xrd characterization of hydroxyapatite from new natural sources. *Spectrochim. Acta Part A-Mol. Biomol. Spectrosc.* **2011**, *83*, 56–60. [CrossRef] [PubMed]
93. Meskinfam, M.; Sadjadi, M.; Jazdarreh, H. Synthesis and characterization of surface functionalized nanobiocomposite by nano hydroxyapatite. *Int. J. Mater. Metall. Eng.* **2012**, *6*, 192–195.
94. Aktug, S.L.; Kutbay, I.; Usta, M. Characterization and formation of bioactive hydroxyapatite coating on commercially pure zirconium by micro arc oxidation. *J. Alloy. Compd.* **2017**, *695*, 998–1004. [CrossRef]
95. Müller, L.; Müller, F.A. Preparation of SBF with different content and its influence on the composition of biomimetic apatites. *Acta Biomater.* **2006**, *2*, 181–189. [CrossRef] [PubMed]
96. Li, H.; Khor, K.A.; Cheang, P. Properties of heat-treated calcium phosphate coatings deposited by high-velocity oxy-fuel (hvof) spray. *Biomaterials* **2002**, *23*, 2105–2112. [CrossRef]
97. Song, W.H.; Jun, Y.K.; Han, Y.; Hong, S.H. Biomimetic apatite coatings on micro-arc oxidized titania. *Biomaterials* **2004**, *25*, 3341–3349. [CrossRef]
98. Li, P.J.; Kangasniemi, I.; Degroot, K.; Kokubo, T. Bonelike hydroxyapatite induction by a gel-derived titania on a titanium substrate. *J. Am. Ceram. Soc.* **1994**, *77*, 1307–1312. [CrossRef]
99. Hu, C.H.; Xia, M.S. Adsorption and antibacterial effect of copper-exchanged montmorillonite on *escherichia coli* k-88. *Appl. Clay Sci.* **2006**, *31*, 180–184. [CrossRef]
100. Ohsumi, Y.; Kitamoto, K.; Anraku, Y. Changes induced in the permeability barrier of the yeast plasma-membrane by cupric ion. *J. Bacteriol.* **1988**, *170*, 2676–2682. [CrossRef]
101. Dan, Z.G.; Ni, H.W.; Xu, B.F.; Xiong, J.; Xiong, P.Y. Microstructure and antibacterial properties of AISI 420 stainless steel implanted by copper ions. *Thin Solid Films* **2005**, *492*, 93–100. [CrossRef]
102. Theivasanthi, T.; Alagar, M. Studies of copper nanoparticles effects on micro-organisms. *Ann. Biol. Res* **2011**, *2*, 368–373.
103. Ramyadevi, J.; Jeyasubramanian, K.; Marikani, A.; Rajakumar, G.; Rahuman, A.A. Synthesis and antimicrobial activity of copper nanoparticles. *Mater. Lett.* **2012**, *71*, 114–116. [CrossRef]
104. Trapalis, C.C.; Kokkoris, M.; Perdikakis, G.; Kordas, G. Study of antibacterial composite Cu/SiO$_2$ thin coatings. *J. Sol-Gel Sci. Technol.* **2003**, *26*, 1213–1218. [CrossRef]
105. Wang, Q.; Larese-Casanova, P.; Webster, T.J. Inhibition of various gram-positive and gram-negative bacteria growth on selenium nanoparticle coated paper towels. *Int. J. Nanomed.* **2015**, *10*, 2885–2894.

© 2018 by the author. Licensee MDPI, Basel, Switzerland. This article is an open access article distributed under the terms and conditions of the Creative Commons Attribution (CC BY) license (http://creativecommons.org/licenses/by/4.0/).

Article

Plasma Electrolytic Oxidation of Titanium in H$_2$SO$_4$–H$_3$PO$_4$ Mixtures

Bernd Engelkamp [1,*], Björn Fischer [2] and Klaus Schierbaum [1]

[1] Abteilung für Materialwissenschaft, Institut für Experimentelle Physik der kondensierten Materie, Heinrich-Heine-Universität Düsseldorf, Universitätsstraße 1, 40225 Düsseldorf, Germany; klaus.schierbaum@hhu.de

[2] Raman Spectroscopic Services, FISCHER GmbH, Necklenbroicher Str. 22, 40667 Meerbusch, Germany; fischer@ramanservice.de

* Correspondence: bernd.engelkamp@hhu.de; Tel.: +49-211-81-15256

Received: 19 December 2019; Accepted: 28 January 2020; Published: 30 January 2020

Abstract: Oxide layers on titanium foils were produced by galvanostatically controlled plasma electrolytic oxidation in 12.9 M sulfuric acid with small amounts of phosphoric acid added up to a 3% mole fraction. In pure sulfuric acid, the oxide layer is distinctly modified by plasma discharges. As the time of the process increases, rough surfaces with typical circular pores evolve. The predominant crystal phase of the titanium dioxide material is rutile. With the addition of phosphoric acid, discharge effects become less pronounced, and the predominant crystal phase changes to anatase. Furthermore, the oxide layer thickness and mass gain both increase. Already small amounts of phosphoric acid induce these effects. Our findings suggest that anions of phosphoric acid preferentially adsorb to the anodic area and suppress plasma discharges, and conventional anodization is promoted. The process was systematically investigated at different stages, and voltage and oxide formation efficiency were determined. Oxide surfaces and their cross-sections were studied by scanning electron microscopy and energy-dispersive X-ray spectroscopy. The phase composition was determined by X-ray diffraction and confocal Raman microscopy.

Keywords: titanium dioxide; plasma electrolytic oxidation; anatase and rutile

1. Introduction

The oxide layer of metals such as titanium can be tailored for specific applications. The most common technique used to artificially grow a passive layer is anodic oxidation, in which moderate voltages promote a denser and thicker oxide layer compared with the naturally formed oxide. With the increasing scope of applications, new demands on materials have evolved. To meet these needs, researchers have developed new techniques from classical anodic oxidation. One derived technique is plasma electrolytic oxidation (PEO), in which the applied voltage exceeds a critical point and causes the initial oxide layer to reform by characteristic breakdowns, which often induce plasma conditions. The complex interplay between chemical, electrochemical, and thermodynamic reactions creates unique oxide layers and enables versatile changes in these layers by slightly changing the process parameters. Therefore, PEO-treated titanium can be used for a variety of applications, such as biomedical prostheses, automotive components, and photocatalytic devices [1].

Furthermore, PEO-treated titanium has recently been used as a gas sensor material at room temperature for various gases [2]. In this case, H$_2$SO$_4$ at an exceptionally high concentration of 12.9 M is used as an electrolyte and leads to a characteristic porous oxide structure with a layer thickness of approximately 5.5 µm. The morphology and thickness suggest a high surface-to-bulk ratio, which is beneficial for gas sensor technology. In general, breakdowns during PEO promote the formation of crystalline titanium dioxide phases, namely, anatase and the high-temperature phase rutile [3–5].

Both phases differ in significant properties (e.g., band-gap energy and electron–hole recombination rate) for potential application as a gas sensor material [6]. When investigating the effect of the crystal phase composition on the gas–oxide interaction, the ability to systematically control the rutile to anatase fraction is desirable. In 12.9 M H_2SO_4, the dominant crystal phase in the oxide is rutile, while the anatase fraction dominates in lower concentrations [7]. However, lower concentrations adversely affect the oxide layer by decreasing its thickness and porosity.

In recent studies on PEO for medical applications, electrolytes with phosphoric acid (H_3PO_4) have been frequently used, and oxide layers of comparable porosity and thickness can be formed [8–10]. Anatase is the dominant phase in these layers, and rutile is almost non-existent. When used as an electrolyte, phosphoric acid not only affects the phase composition but also drastically changes the outcome of the PEO process. The combination of both electrolytes provides an interesting approach to tailoring the properties of the oxide scale. Since the ratio of the two compounds in the mixture is critical for obtaining specific properties, we explored the effect of small amounts of H_3PO_4 in concentrated H_2SO_4.

Our PEO experiment is based on a galvanostatic DC operation mode. The resulting constant current offers a simple method of treating and evaluating samples for a systematic PEO study. For instance, it can be split into several contributions and classified into ionic and electronic currents [8,11]. The ionic current reflects the migration of ions in the oxide layer and is the driving force in conventional anodic oxidation. The electronic current is mainly induced by breakdowns. In the course of the PEO process, a transition from ionic to electronic current can be observed.

Starting with concentrated sulfuric acid (12.9 M) as an electrolyte, we investigated the impact of adding H_3PO_4 at molar fractions of 1% and 3%. Before and after PEO, the samples were analyzed for weight gain with a microbalance. Scanning electron microscopy and X-ray diffraction were used to systematically study oxide surfaces and cross sections at different stages of the process. We derived information about the phase distribution in the oxide layer from confocal Raman microscopy of cross sections, and elemental composition was investigated by energy-dispersive X-ray spectroscopy.

2. Materials and Methods

Samples (surface area of approx. 3.634 cm^2) were cut from titanium foil (thermally annealed, 99.6% purity; 125 µm thickness) by means of a laser (PowerLine F30, ROFIN-SINAR Laser GmbH, Hamburg, Germany). The samples were cleaned ultrasonically for approximately 10 min in acetone and 10 min in deionized water. The electrolytic cell consisted of a glass vessel with an integrated glass shell, which enabled the temperature regulation of the electrolyte by pumping cold thermal fluid through the shell. The thermal fluid temperature was kept constant at 15 °C with a recirculating cooler (FL1201, JULABO GmbH, Seelbach, Germany). The reaction chamber was filled with 114 ± 5 mL of the electrolyte. The electrolyte was based on 12.9 M H_2SO_4 (75 wt%). It was enriched with 25 wt% H_3PO_4, which resulted in molar fractions of $n(H_3PO_4)/n(H_2SO_4)$ = 0%, 1%, and 3%. Specifically, the last two fractions corresponded to $c_{1\%}(H_3PO_4)$ = 0.1 M plus $c_{1\%}(H_2SO_4)$ = 12.3 M and $c_{3\%}(H_3PO_4)$ = 0.3 M plus $c_{3\%}(H_2SO_4)$ = 11.4 M. A magnetic stirrer prevented spatial temperature differences in the electrolyte and reduced the disturbance of gas accumulations on the electrode surfaces. The electrolytic system was completed by the titanium sample as the anode and a graphite rod as the cathode (area of immersion of 7.38 ± 0.6 cm^2) at a distance of 23 ± 4 mm.

A constant current density of 55 mA/cm^2 was applied by using a highly stable current power supply (FUG MCP 350-350). Voltage and current were adjusted and recorded in 250 ms intervals using an in-house developed LabVIEW program. After treatment, the sample was rinsed in deionized water and dried in air. The weight of the sample was measured before and after the process with an analytical balance (ABT 120-5DM, Kern und Sohn GmbH, Germany) with a repeatability of 0.02 mg.

The microstructure of the oxide layer was investigated by field emission scanning electron microscopy (SEM; JSM-7500F, JEOL Ltd., Tokyo, Japan). Surface images of secondary electrons were captured with 5 kV excitation. Cross sections were prepared by argon ion milling (Cross Section Polisher IB-09010CP, JEOL Ltd., Tokyo, Japan). For energy-dispersive X-ray spectroscopy (EDX), images were created with 15 kV excitation energy and detected with an XFlash Detector 5030 (Bruker AXS GmbH, Karlsruhe, Germany). Quantitative results of elemental composition were obtained by averaging over at least three comparable sections to minimize local fluctuations.

The phase composition was determined by X-ray diffraction (XRD). Diffraction data were collected on a Bruker D2 Phaser diffractometer with Cu-Kα radiation (λ = 1.54184 Å, 30 kV, 10 mA) in Bragg–Brentano geometry and a LYNXEYE 1D detector. XRD patterns were measured with a flat silicon, low-background rotating sample holder (5.0 min^{-1}) with 24.5° < 2θ < 29.5°, a scan speed of 2 s/step, and a step size of approximately 0.024°.

Raman measurements were performed with a confocal Raman microscope alpha300 R (WITec GmbH, Ulm, Germany). A fiber-coupled single-mode DPSS laser with an excitation wavelength of 532 nm was used. The laser power applied to the sample was set to 20 mW. A Zeiss EC Epiplan-Neofluar DIC 100x/0.9 NA was selected as the microscope objective, and the samples were scanned with a step size of 200 nm. In this way, a spatial resolution of about 300 nm could be achieved. The system also featured real-time laser profilometry, so the sample surface remained in the focal plane during the entire measurement period. The spectrometer used was a WITec UHTS 300 combined with an Andor iDus Deep Depletion CCD detector, which was cooled to −60 °C. The Raman scattered light was spectrally dispersed by a reflection grating with 1200 mm^{-1}. The average spectral resolution was about 2 cm^{-2}/pixel. The software WITec FIVE version 5.2.4.81 was used to evaluate the measurement data and create Raman images, including cosmic ray removal and background subtraction by the implemented shape function.

3. Results

3.1. Voltage Response and Mass Change

The voltage response of the PEO process in 12.9 M H_2SO_4, as shown in Figure 1a, reveals information about the current character. The ratio between electronic and ionic currents varies during the PEO process. In the beginning, the electric field was insufficient to cause electric breakdowns, and conventional anodic oxidation occurs. The ionic current predominated and promoted the formation of a dielectric oxide layer. Consequently, the cell resistance increased. The constant current was sustained by the rapid increase in applied voltage. Above a critical voltage, electrical breakdowns become visible by electroluminescence. The charge transfer by breakdowns represents an electronic current and is energetically favored compared with ion migration. The electronic current increasingly contributes to the total current. Eventually, a linear voltage region is reached, which indicates a mainly electronic current character. This linear stage is also known as the microarc stage [9,12].

The impact of breakdowns on oxide formation was further investigated by terminating the process at different charge densities and determining the mass change, presumably due to oxide formation, with an analytical balance. Figure 1b presents the mass change with varying charge densities. The slope of the polynomial fit represents the mass change per transferred charge, i.e., the oxide formation efficiency [13]. The efficiency for 0% H_3PO_4 was positive until 8.4 C/cm^2, after which it was negative. This turning point corresponded to the beginning of the linear microarc stage. This indicates that breakdowns during the microarc stage in concentrated sulfuric acid cause mass loss of the oxide. The initial passivation by breakdowns transforms into a destructive reforming with combined mass loss.

The progress drastically changed by adding small amounts of H_3PO_4 to concentrated sulfuric acid. The voltage response for n(H_3PO_4)/n(H_2SO_4) = 1% in Figure 1a already differed from the voltage response in pure H_2SO_4. The transition from ionic to electronic current was also observable by the subsequent decreasing voltage rate. However, it was less pronounced. The interruption of the steep increase between 1.5 and 7 C/cm^2 is due to the previously reported transition from a grooved morphology to a porous morphology [13,14]. The mass loss in Figure 1b was suppressed compared with samples prepared in pure H_2SO_4. The efficiency was positive until 19.4 C/cm^2. Shortly after the efficiency changes to negative values, the microarc region started, and no noticeable voltage gain occurred. Hence, the applied voltage was sufficient to sustain the defined current density, even though the oxide formation efficiency in this region (Figure 1b) was negative. Therefore, a steady state between repassivation of the dielectric layer and its destruction by breakdowns can be assumed.

With 3% phosphoric acid, the trend continued more drastically. Higher voltages, even above 200 V, were necessary to sustain the given current density. The transition from ionic to electronic currents was completed even later, while the mass change was always positive. At approximately 55.8 C/cm^2, the efficiency changed from positive to negative values, and the microarc region started. However, above approximately 70 C/cm^2, inhomogeneities were visible on the oxide surface and restricted the process in 3% H$_3$PO$_4$ at higher charge densities.

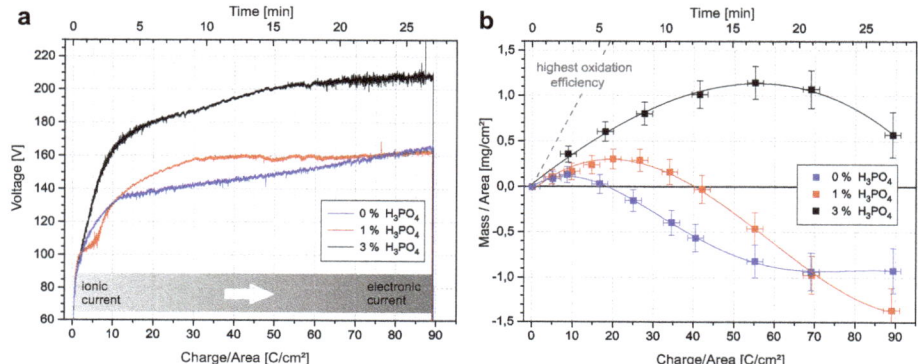

Figure 1. Process information at different charge densities for 0%, 1%, and 3% additional H$_3$PO$_4$ in 12.9 M H$_2$SO$_4$: (**a**) voltage response and (**b**) mass change with interpolation and marked highest oxidation efficiency.

3.2. XRD Investigation

The crystallographic structure of the oxide was investigated by X-ray diffraction (XRD). The weight fraction of rutile can be estimated for each sample by using the (101) reflection of anatase and the (110) reflection of rutile [7,15]. Figure 2 presents the calculated rutile fraction for different charge densities. The rutile fraction in the sample prepared with 0% H$_3$PO$_4$ increased continuously with the charge density until the sample was almost exclusively rutile. The titanium reflections of the titanium substrate decreased continuously as a result of the growing oxide layer [7]. The transition was similar in 1% H$_3$PO$_4$, but it was less pronounced. At a low charge density, mainly anatase was present, and the rutile fraction increased with increasing charge density. However, the maximum value and the slope of the fitted curve are smaller. For 3% H$_3$PO$_4$, our findings were remarkably different, and the typical increase in the rutile to anatase fraction was no longer identified. The fraction was below 11% for any charge density.

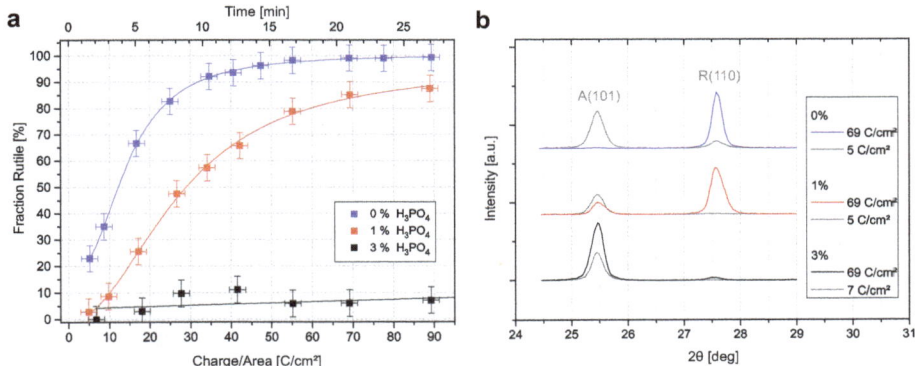

Figure 2. (a) Rutile to anatase fractions (with interpolation) versus the charge density for samples prepared with 0%, 1%, and 3% additional H_3PO_4 in 12.9 M H_2SO_4. The fractions are derived from XRD intensities. (b) Representative diffractograms of chosen samples with high and low current densities.

3.3. SEM Surface Images

The scanning electron microscope (SEM) images of surfaces in Figure 3 demonstrate the modification of surfaces resulting from variations in the transferred charge and the amount of H_3PO_4. In the left column, the surfaces of samples prepared with 0% H_3PO_4 are shown. Since the breakdown voltage was already exceeded for 9 C/cm^2, circular sinkholes of former discharge channels, i.e., micropores, were clearly visible. However, the even distribution caused the surface to appear regular and flat. With increasing charge transfer, the surface became rougher. At 41 C/cm^2, some minor plateaus and some cracks were noticeable. For 69 C/cm^2, plateaus and cracks were clearly visible and impaired the circular shape of pores.

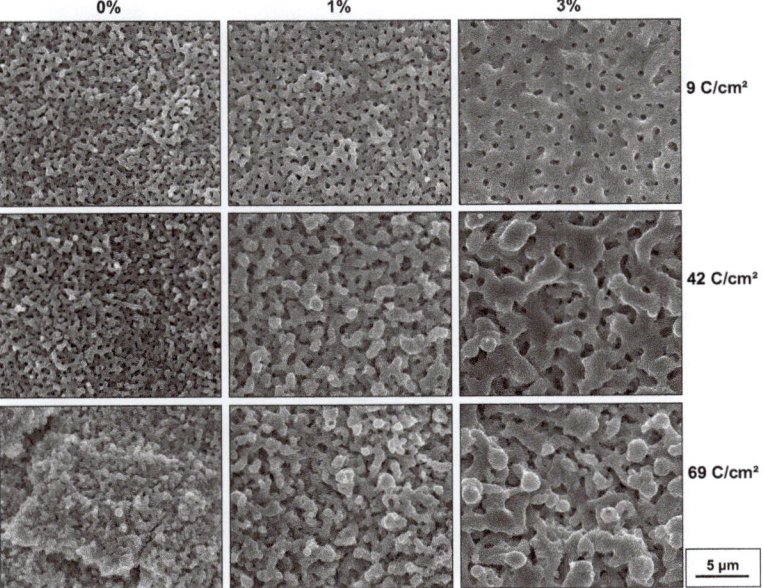

Figure 3. SEM images of sample surfaces produced in H_2SO_4 with the addition of 0%, 1%, or 3% H_3PO_4. Samples produced with similar charge densities (± 1.0 C/cm^2) are presented in the same row.

In the second column, the surfaces of the samples prepared in 1% H_3PO_4 are shown. The porous structure remained apparent. However, the pore size increased, and the pore density decreased. As observed previously, the pore network dissolved at a higher charge density, and different levels of depth evolved. In the right column, the surfaces of samples prepared in 3% H_3PO_4 are presented. The pore size further increased, and the density further decreased. The destructive character of the breakdowns can be observed, although it was less pronounced compared with the samples prepared in 0% or 1% H_3PO_4.

3.4. SEM Cross Sections

The result of our SEM cross section investigation in Figure 4 shows the depth profile of the oxides. The total thickness distinctly increased with the fraction of phosphoric acid in the electrolyte. Values were approximately 5.0 µm for 0%, 7.3 µm for 1%, and 10.9 µm for 3% (±0.5 µm). Different layers could be distinguished in the oxide layer because they abruptly changed in morphology or elemental composition. Two layers were as described for PEO in sulfuric acid: a compact layer beside the titanium substrate with a relatively small thickness and a porous layer with a major contribution to the total thickness [7,16]. Furthermore, in Figure 4c, a smooth area in the near the surface was clearly distinguishable from the porous layer below.

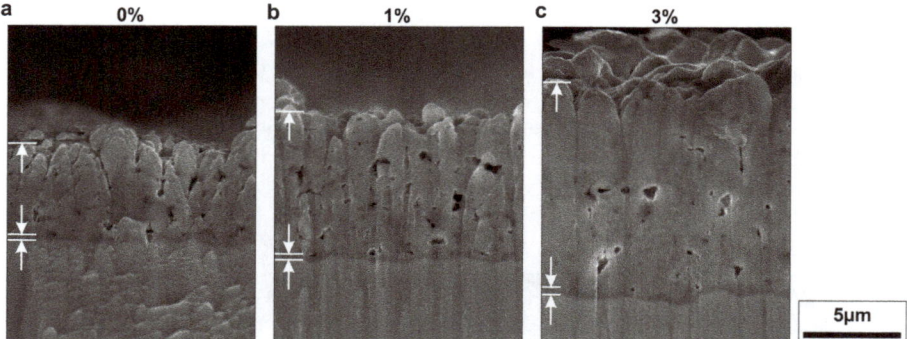

Figure 4. SEM images of oxide cross sections produced in H_2SO_4 with the addition of 0% (**a**), 1% (**b**), and 3% (**c**) H_3PO_4. The charge density in all plasma electrolyte oxidation (PEO) processes is 41.3 ± 1.0 C/cm^{-2}.

3.5. EDX Images

While the compact layer is rather difficult to resolve in the SEM cross section image, it is clearly identifiable by its elemental composition. This is shown in Figure 5 by the energy-dispersive X-ray spectroscopy (EDX) images. Remarkably, the compact layer in all samples exhibited an increased sulfur concentration of approximately 1.1 ± 0.3 at % (while it was below 0.3 ± 0.3 at % in the remaining oxide). In samples prepared with H_3PO_4, an increased phosphor concentration throughout the oxide could be detected. For 3% H_3PO_4, the value is approximately 1.9 ± 0.3 at %. In comparison, the phosphor concentration in a sample produced in 1% H_3PO_4 amounted to 0.6 ± 0.3 at %. Another feature was detectable in the sample produced in 3% H_3PO_4 and less pronounced in the sample produced in 0% H_3PO_4: a smooth area near the surface was observed in several samples and independent of the electrolyte composition. The EDX analysis in Figures 5a,c reveals that the region exhibited an increased titanium concentration, while the oxygen and phosphor concentration decreased.

3.6. Confocal Raman Microscopy

Since a common drawback of XRD phase analysis is the low spatial resolution, we additionally performed confocal Raman microscopy. Figure 6 shows false-color images, which are derived from Raman microscopy of cross sections prepared with 41.3 ± 1.0 C/cm^{-2} in mixtures with 0%, 1%, and 3% additional H_3PO_4. The surrounding background image is the result of light microscopy. The different colors in the false-color images indicate the types of Raman spectra, which are presented below. The anatase phase is recognized by an intense E_g mode around 147 cm^{-1} [17]. Additionally, Raman-active modes derived from anatase around 395 cm^{-1} (B_{1g}), 515 cm^{-1} (B_{1g} and A_{1g}), and 637 cm^{-1} (E_g) are used for classification [7]. The rutile phase can be identified by two Raman modes around 442 cm^{-1} (E_g) and 605 cm^{-1} (A_{1g}) [7].

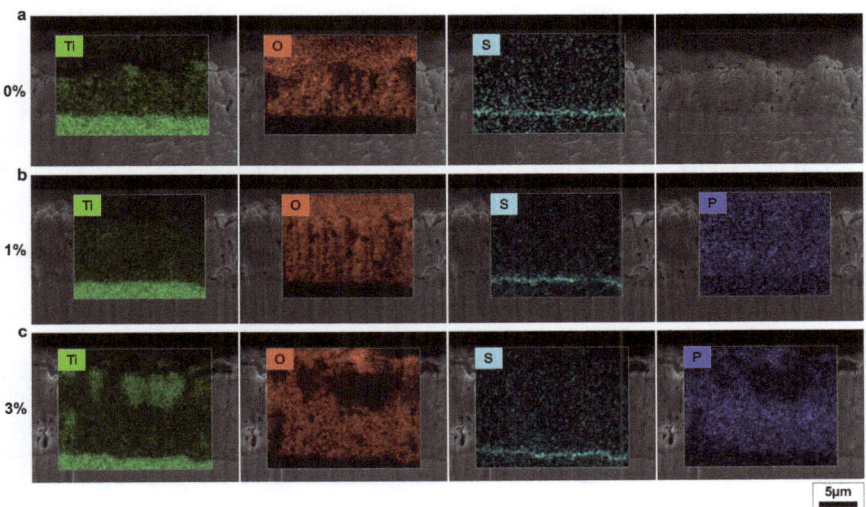

Figure 5. Energy-dispersive X-ray spectroscopy (EDX) images of oxide cross sections produced in 12.9 M H_2SO_4 with 0% (**a**), 1% (**b**), and 3% (**c**) H_3PO_4. The transferred charge density is around 41.3 ± 1.0 C/cm^{-2} for each sample. The color intensity in the analyzed segment is only comparable to other elemental maps for the same cross section.

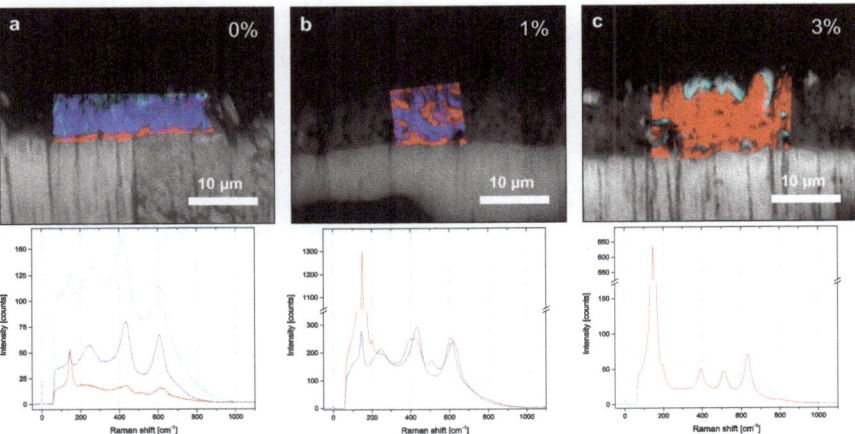

Figure 6. False-color images derived from confocal Raman microscopy with underlying light microscopy images of three cross sections. The transferred charge density is around 41.3 ± 1.0 C/cm^{-2} for all cross sections, while the fraction of H_3PO_4 in the electrolyte changes, i.e., 0% (**a**), 1% (**b**), and 3% (**c**). Red represents anatase, blue represents rutile, and cyan corresponds to a signal with large background. The integration time of each single spectrum varies, with 0.2 s in 0%, 0.1 s in 1%, and 0.05 s in 3%. Averaged spectra from the corresponding color-marked area are shown below each cross section.

Without H_3PO_4, the porous layer mainly exhibited rutile, while a distinct anatase fraction could be identified from the spectra of the compact layer. For 1% H_3PO_4, the colored areas, which indicate anatase and rutile, were similar in size and homogeneously distributed. For 3%, the intensities of the rutile modes in the spectra vanished. The entire oxide mainly exhibited anatase. For 0% and 3%, small areas near the surface were marked with cyan and fitted the previously mentioned smooth area in our SEM and EDX images. The corresponding cyan spectra resembled the spectra from rutile. However, it clearly differed by a frequency shift and an increased background signal, which may have originated from near-surface groups.

4. Discussion

Our results demonstrate that adding phosphoric acid to concentrated sulfuric acid as an electrolyte has a drastic influence on oxide formation in PEO. With 1% or 3% additional phosphoric acid, higher potentials are necessary to sustain the constant current density. This indicates an enhanced dielectric layer in terms of electrical resistance, with either increased thickness or higher electrical resistivity. The SEM cross sections confirm an increasing thickness. The promoted oxidation is associated with the mass gain in Figure 1b, which is mainly negative in 12.9 M H_2SO_4, partly positive with 1% H_3PO_4, and positive for all investigated charge densities with 3% H_3PO_4.

While oxide formation is promoted, breakdown effects are inhibited by adding H_3PO_4, which results in diminished breakdowns [10,14]. One indicator is the suppressed surface destruction, as shown in Figure 3. Breakdowns may cause plasma oxidation; however, almost all electric energy is used for ionization, water vaporization, joule heating, and gas evolution, especially in the microarc region [18]. As a consequence, plateaus and cracks evolve on the surface. After adding H_3PO_4 to the electrolyte, these effects decrease. Closely related is the mentioned mass loss in Figure 1b. It is assumed that the mass loss is promoted by breakdowns. Since the mass change becomes positive with additional H_3PO_4, lower breakdown intensity is expected.

Furthermore, reduced breakdown intensity can be identified from the phase composition. Rutile is a high-temperature modification and formed from anatase in an irreversible, time-dependent transformation [6]. In PEO, the energy for transformation is brought into the system by electrical breakdowns. When breakdowns diminish in the process, transformation is reduced or even disabled [7]. Results derived from the XRD analysis in Figure 2 show that the rutile fraction decreases with an increasing amount of H_3PO_4. Confocal Raman microscopy data, as presented in Figure 6, confirm this observation. Our results suggest that the energy liberated by breakdowns decreases with an increasing amount of H_3PO_4. With even higher concentrations of H_3PO_4, it is assumed that the crystallization is further inhibited and that even the anatase fraction is reduced [19].

In our discussion of the impact of additional H_3PO_4, we must consider that the concentrations of both acids decrease if one mixes appropriate amounts of H_2SO_4 (75 wt%) and H_3PO_4 (25 wt%). For example, when adjusting a fraction of $n(H_3PO_4)/n(H_2SO_4) = 3\%$, the resulting concentration of H_2SO_4 is 11.4 M, while the concentration of H_3PO_4 is 0.3 M. To understand the dilution influence of H_2SO_4, we conceived a PEO experiment with 11.4 M H_2SO_4. As a result, no significant change in the breakdown character was observed. In conclusion, the additional H_3PO_4 causes the drastic change observed in the PEO process.

To rationalize our findings, we correlated the impact of additional H_3PO_4 in 12.9 M H_2SO_4 to the corrosion of titanium in the electrolyte. In pure H_2SO_4 and the absence of an applied potential, the maximum corrosion rate is between 12.5 [20] and 13.7 M [21]. With an applied concentration of 12.9 M, a strong chemical etching can therefore be expected. In contrast, a simple corrosion experiment without applied potential in 12.9 M H_2SO_4 with 3% H_3PO_4 clearly demonstrates that the corrosion rate drastically decreases. A possible explanation is that solvated H_3PO_4 anions preferentially adsorb onto the anodic area. A similar effect is known to occur for corrosion inhibitors in H_2SO_4 [20]. As a consequence of the preferential adsorption, the active area is blocked for the more reactive H_2SO_4 anions.

The concept of the described corrosion behavior can be applied to our PEO experiment. In 12.9 M, the breakdowns are exceptionally strong, whereas the breakdowns are suppressed in a mixture with 3% H_3PO_4. We assume that, during PEO, the preferential adsorption of H_3PO_4 anions blocks the anodic surface for H_2SO_4 anions, which tend to favor breakdowns instead of ion migration. Hence, destructive breakdowns and the related corrosion are suppressed, which directly reduces the mass loss in the process, as confirmed by our mass investigation in Figure 1b. Leach and Sidgwick proposed that the different behaviors of SO_4^{2-} and PO_4^{3-} are the result of their different molecular charges [22], which is plausible since the electrical field in PEO is exceptionally high.

Furthermore, the high H_3PO_4 anion adsorption to the anodic surface leads to the preferred incorporation of H_3PO_4 anions. This is in accordance with our EDX investigation, which reveals an increased concentration of phosphor species compared with the low sulfur concentration in the oxide. Similarly, earlier studies have confirmed that phosphorous or phosphate ions penetrate more easily through the titanium oxide layer during anodization compared with sulfur or sulfate ions [19]. Certainly, these phosphorous species contribute to the high mass gain in Figure 1b for electrolytes with additional H_3PO_4.

With the favored incorporation of H_3PO_4 anions, enhanced ion migration can be assumed, which increases the fraction of the ionic current. Because of the constant current mode in our process, the total current density always remains constant. When ionic migration is promoted, the total current in the process comprises less electronic current, i.e., breakdowns. Therefore, the additional H_3PO_4 reinforces the inhibition of breakdowns.

The reduced electronic current has a major impact on the oxide layer. According to the literature, two effects limit the lifetime of a discharge. On the one hand, the expansion of gases in the channel lead to cooling, and the plasma collapses [23]. On the other hand, the gas forms a bubble on top of the channel; thus, it increases the electrical resistivity of the channel and terminates the plasma [4,18]. The formation of a new plasma is prevented while the gas is inside and above the channel. When the gas escapes, the electrolyte fills the void [3]. As a consequence, the next discharge is created in the same channel. This hypothesis predicts that the current per channel per discharge is limited by the lifetime of the plasma channel.

According to our results, more charge compensation in the form of breakdowns occurs in pure H_2SO_4 compared with mixtures containing 1% or 3% H_3PO_4. As a consequence of the limited charge transfer per discharge in a single channel, more channels are necessary for higher charge compensation. Hence, the pore density increases compared with surfaces prepared in mixtures with 1% or 3% H_3PO_4, as shown in the SEM images in Figure 3. On the other hand, fewer channels for charge compensation by breakdowns are necessary in mixtures with H_3PO_4. Hence, the pore density decreases. However, the pore size increases in mixtures with H_3PO_4 because of recurring breakdowns in identical channels.

It should be noted that anatase is the predominant phase in the compact layer for all investigated samples. Previous studies confirm our conclusion that the thin compact layer is composed of nanocrystalline anatase [8,24]. The dense compact layer likely evolves from a temperature gradient [9]. The thermal mass of the titanium substrate enables the rapid cooling of the plasma. Hence, the time is too short and the temperature is too low for the phase transformation to rutile for an anatase-to-rutile transformation, and therefore, only a nanocrystalline anatase structure evolves. Also remarkable is the increased sulfur concentration in the compact layer, as seen in Figure 5. This may result from the electrolyte becoming trapped in interfacial nanopores during the fast cooling [14].

Another notable feature is apparent in the cross sections of several samples, which can be identified in our study in three ways. First, SEM cross section images, especially in Figure 4c, reveal a rather smooth morphology between the porous layer and the surface. Second, the EDX cross sections in Figures 5a,c show a different elemental composition compared with the porous layer in the previously mentioned region. Third, the Raman investigations in Figures 6a,c reveal a distinct change in the phase composition near the surface. We assume that this region originates from molten titanium, which is ejected from the channels and rapidly cools down in the vicinity of the electrolyte [25].

Because of quenching, oxide formation is suppressed, and consequently, oxide stoichiometry is not achieved. For this reason, we expect a predominant amorphous structure with small contributions of titanium dioxide phases.

5. Conclusions

Oxide coatings on titanium were produced in a galvanostatically controlled PEO process with a constant current density. With a concentrated 12.9 M H_2SO_4 electrolyte as the starting material, small amounts of H_3PO_4 were added to investigate the impact on oxide formation. Pure 12.9 M H_2SO_4 is highly suitable for promoting breakdowns. With a higher charge density, breakdowns cause a destructive reforming and induce the phase transition from predominantly anatase to almost entirely rutile. Upon reaching 35 C/cm^2, the rutile to anatase fraction is over 90%.

By adding small amounts of H_3PO_4, i.e., 1% or 3%, breakdown effects are drastically reduced. The rutile fraction in the process does not exceed 11%, even for charge transfers as high as 89 C/cm^2. The drastic change is explained by the preferential adsorption of H_3PO_4 anions to the anodic area. Therefore, the H_3PO_4 anions block the surface for more reactive H_2SO_4 anions and suppress breakdowns. The enhanced concentration of H_3PO_4 anions at the surface reinforces their incorporation and, consequently, migration in the oxide layer. Therefore, the ionic current in the process increases, and the electronic current fraction, including breakdowns, decreases. With fewer breakdowns, the destructive reforming diminishes. Moreover, the increase in ion migration promotes oxide formation with increased thickness and mass gain of the oxide layers, which are produced in mixtures with H_3PO_4.

Author Contributions: B.F. designed, conceived, and performed the confocal Raman microscopy; B.E. designed, conceived, and performed the other experiments; B.E. validated and analyzed the results; K.S. supervised the project; B.E. wrote the paper. All authors have read and agreed to the published version of the manuscript.

Funding: We acknowledge support by the Heinrich Heine University Duesseldorf. This research was also supported by Bundesministerium für Wirtschaft und Energie (BMWi) under project no. ZF4185502ZG6.

Acknowledgments: We gratefully thank Denis Netschitailo for supplementary PEO experiments and related discussions.

Conflicts of Interest: The authors declare no conflict of interest.

References

1. Diamanti, M.V.; Del Curto, B.; Pedeferri, M. Anodic oxidation of titanium: From technical aspects to biomedical applications. *J. Appl. Biomater. Biomech.* **2011**, *9*, 55–69. [CrossRef] [PubMed]
2. El Achhab, M.; Schierbaum, K. Gas sensors based on plasma-electrochemically oxidized titanium foils. *J. Sens. Syst.* **2016**, *5*, 273–281. [CrossRef]
3. Clyne, T.W.; Troughton, S.C. A review of recent work on discharge characteristics during plasma electrolytic oxidation of various metals. *Int. Mater. Rev.* **2018**, *64*, 1–36. [CrossRef]
4. Rakoch, A.G.; Khokhlov, V.V.; Bautin, V.A.; Lebedeva, N.A.; Magurova, Y.V.; Bardin, I.V. Model concepts on the mechanism of microarc oxidation of metal materials and the control over this process. *Prot. Met.* **2006**, *42*, 158–169. [CrossRef]
5. Yerokhin, A.; Nie, X.; Leyland, A.; Matthews, A.; Dowey, S. Plasma electrolysis for surface engineering. *Surf. Coat. Technol.* **1999**, *122*, 73–93. [CrossRef]
6. Hanaor, D.A.H.; Sorrell, C.C. Review of the anatase to rutile phase transformation. *J. Mater. Sci.* **2011**, *46*, 855–874. [CrossRef]
7. Engelkamp, B.; El Achhab, M.; Fischer, B.; Kökçam-Demir, I.; Schierbaum, K. Combined Galvanostatic and Potentiostatic Plasma Electrolytic Oxidation of Titanium in Different Concentrations of H_2SO_4. *Metals* **2018**, *8*, 386. [CrossRef]
8. Mortazavi, G.; Jiang, J.; Meletis, E.I. Investigation of the plasma electrolytic oxidation mechanism of titanium. *Appl. Surf. Sci.* **2019**, *488*, 370–382. [CrossRef]

9. Quintero, D.; Galvis, O.; Calderón, J.; Castaño, J.; Echeverría, F. Effect of electrochemical parameters on the formation of anodic films on commercially pure titanium by plasma electrolytic oxidation. *Surf. Coat. Technol.* **2014**, *258*, 1223–1231. [CrossRef]
10. Friedemann, A.; Gesing, T.; Plagemann, P. Electrochemical rutile and anatase formation on {PEO} surfaces. *Surf. Coat. Technol.* **2017**, *315*, 139–149. [CrossRef]
11. Schultze, J.; Lohrengel, M. Stability, reactivity and breakdown of passive films. Problems of recent and future research. *Electrochim. Acta* **2000**, *45*, 2499–2513. [CrossRef]
12. Barati Darband, G.; Aliofkhazraei, M.; Hamghalam, P.; Valizade, N. Plasma electrolytic oxidation of magnesium and its alloys: Mechanism, properties and applications. *J. Magnes. Alloy.* **2017**, *5*, 74–132. [CrossRef]
13. Galvis, O.; Quintero, D.; Castaño, J.; Liu, H.; Thompson, G.; Skeldon, P.; Echeverría, F. Formation of grooved and porous coatings on titanium by plasma electrolytic oxidation in H_2SO_4/H_3PO_4 electrolytes and effects of coating morphology on adhesive bonding. *Surf. Coat. Technol.* **2015**, *269*, 238–249. [CrossRef]
14. Kern, P.; Zinger, O. Purified titanium oxide with novel morphologies upon spark anodization of Ti alloys in mixed H_2SO_4/H_3PO_4 electrolytes. *J. Biomed. Mater. Res.* **2007**, *80A*, 283–296. [CrossRef]
15. Spurr, R.A.; Myers, H. Quantitative analysis of anatase-rutile mixtures with an X-ray diffractometer. *Anal. Chem.* **1957**, *29*, 760–762. [CrossRef]
16. Fadl-Allah, S.A.; El-Sherief, R.M.; Badawy, W.A. Electrochemical formation and characterization of porous titania (TiO_2) films on Ti. *J. Appl. Electrochem.* **2008**, *38*, 1459–1466. [CrossRef]
17. Balachandran, U.; Eror, N.G. Raman spectra of titanium dioxide. *J. Solid State Chem.* **1982**, *42*, 276–282. [CrossRef]
18. Troughton, S.C.; Nominé, A.; Nominé, A.V.; Henrion, G.; Clyne, T.W. Synchronised electrical monitoring and high speed video of bubble growth associated with individual discharges during plasma electrolytic oxidation. *Appl. Surf. Sci.* **2015**, *359*, 405–411. [CrossRef]
19. Lee, J.H.; Kim, S.E.; Kim, Y.J.; Chi, C.S.; Oh, H.J. Effects of microstructure of anodic titania on the formation of bioactive compounds. *Mater. Chem. Phys.* **2006**, *98*, 39–43. [CrossRef]
20. Abdel Hady, Z.; Pagetti, J. Anodic behaviour of titanium in concentrated sulphuric acid solutions. Influence of some oxidizing inhibitors. *J. Appl. Electrochem.* **1976**, *6*, 333–338. [CrossRef]
21. Prando, D.; Brenna, A.; Diamanti, M.V.; Beretta, S.; Bolzoni, F.; Ormellese, M.; Pedeferri, M. Corrosion of titanium: Part 2: Effects of surface treatments. *J. Appl. Biomater. Funct. Mater.* **2018**, *16*, 3–13. [CrossRef]
22. Leach, J.S.L.; Sidgwick, D.H. Anodix oxidation of Titanium. *Metall. Corrosion Proc.* **1981**, *1*, 82–85.
23. Klapkiv, M.D. Simulation of synthesis of oxide-ceramic coatings in discharge channels of a metal-electrolyte system. *Mater. Sci.* **1999**, *35*, 279–283. [CrossRef]
24. El Achhab, M.; Erbe, A.; Koschek, G.; Hamouich, R.; Schierbaum, K. A microstructural study of the structure of plasma electrolytically oxidized titanium foils. *Appl. Phys. A* **2014**, *116*, 2039–2044. [CrossRef]
25. Mohedano, M.; Lu, X.; Matykina, E.; Blawert, C.; Arrabal, R.; Zheludkevich, M. Plasma electrolytic oxidation (PEO) of metals and alloys. *Encycl. Interfacial Chem.* **2018**, *6*, 423–438.

© 2020 by the authors. Licensee MDPI, Basel, Switzerland. This article is an open access article distributed under the terms and conditions of the Creative Commons Attribution (CC BY) license (http://creativecommons.org/licenses/by/4.0/).

Article

LDH Post-Treatment of Flash PEO Coatings

Rubén del Olmo [1,*], Marta Mohedano [1], Beatriz Mingo [2], Raúl Arrabal [1] and Endzhe Matykina [1]

1. Departamento de Ciencia de Materiales, Facultad de Ciencias Químicas, Universidad Complutense, 28040 Madrid, Spain; mmohedano@ucm.es (M.M.); rarrabal@ucm.es (R.A.); ematykin@ucm.es (E.M.)
2. School of Materials, University of Manchester, Oxford Road, Manchester M13 9PL, UK; beatriz.mingo@manchester.ac.uk
* Correspondence: rubandom@ucm.es; Tel.: +34-913-945-227

Received: 25 April 2019; Accepted: 28 May 2019; Published: 30 May 2019

Abstract: This work investigates environmentally friendly alternatives to toxic and carcinogenic Cr (VI)-based surface treatments for aluminium alloys. It is focused on multifunctional thin or flash plasma electrolytic oxidation (PEO)-layered double hydroxides (LDH) coatings. Three PEO coatings developed under a current-controlled mode based on aluminate, silicate and phosphate were selected from 31 processes (with different combinations of electrolytes, electrical conditions and time) according to corrosive behavior and energy consumption. In situ Zn-Al LDH was optimized in terms of chemical composition and exposure time on the bulk material, then applied to the selected PEO coatings. The structure, morphology and composition of PEO coatings with and without Zn-Al-LDH were characterized using XRD, SEM and EDS. Thicker and more porous PEO coatings revealed higher amounts of LDH flakes on their surfaces. The corrosive behavior of the coatings was studied by electrochemical impedance spectroscopy (EIS). The corrosion resistance was enhanced considerably after the PEO coatings formation in comparison with bulk material. Corrosion resistance was not affected after the LDH treatment, which can be considered as a first step in achieving active protection systems by posterior incorporation of green corrosion inhibitors.

Keywords: PEO; LDH; active protection; corrosion; aluminium

1. Introduction

Plasma electrolytic oxidation (PEO) is a plasma-assisted electrochemical surface treatment characterized by the utilization of high voltages (100–600 V) in alkaline electrolytes to produce ceramic-like coatings on light alloys such as aluminium [1], magnesium [2] and titanium [3]. This technology may become a potential alternative to conventional toxic and highly carcinogenic chromic acid anodizing (CAA) for niche applications [4–6].

The PEO process can be conducted under direct current (DC) [7], alternate current (AC) [8], unipolar [9] or bipolar pulsed regimes [10] involving polarization of the alloy to voltages above the dielectric breakdown of the oxide. This results in the formation of multiple short-lived microdischarges on the metallic surface that trigger the formation of highly stable ceramic phases [1,11,12]. The resulting PEO coatings present high hardness, thermal stability and adherence to the substrate, which lead to enhanced corrosion and tribological properties. Moreover, the coatings' microstructures and compositions can be tailored as required by controlling electrical parameters during the coating synthesis by electrolyte selection and by the application of pre- and post-treatments [13,14].

However, the costs associated with PEO technology are relatively high, which is mainly due to high energy consumption. General current densities and voltages for the PEO process are in the range of 1.5–15 A·dm^{-2} and 400–500 V for Mg [15], 6–20 A·dm^{-2} and 200–500 V for Ti [16] and 10–60 A·dm^{-2} and 200–400 V for Al [17]. Typical PEO treatment times are within a 15–60 min range.

In the particular case of Al and its alloys, the efficiency of conversion of the anodic charge into the final coating material may be as low as 20%, due to gas generation and dissolution processes [18–20]. Therefore, energy consumption of the process must be reduced in order to make this technology commercially viable for high-volume applications [10,14,20,21]. Strategies to achieve this goal involve waveform design and cell geometry [22], electrolyte design [23–27]) and a pre-anodizing approach [18]. An additional strategy consists of "flash" PEO—processes of a short duration (1–3 min) that produce thin coatings (1–5 μm) suitable and sufficient for many corrosion resistance-demanding applications.

PEO coating properties can be improved even further by incorporating active functionalities. This can be achieved by introducing smart agents into the coatings matrix, which are released in situ, triggered by an external stimulus [23–25]. Coatings based on LDH have increasing potential not only for developing smart functionalities on ceramic materials, but in many other fields, such as biomaterials, due to their low toxicity [23].

LDHs can be expressed by the general formula $[M^{2+}_{1-X}M^{3+}_{X}(OH)_2]^{X+}(A^{m-})_{X/m} \cdot nH_2O]$, where M^{2+} and M^{3+} represent divalent metallic cations (e.g., Mg^{2+}, Zn^{2+}, etc.) and trivalent metallic cations (e.g., Al^{3+}, Co^{3+}, etc.), respectively, and A^{m-} is the interlayer anion (e.g., NO_3^-, Cl^-, etc.) [26]. It consists of uniform flakes that grow roughly vertically on the substrate and act as intelligent nanocontainers that release, in a controlled way, the previously loaded corrosion inhibitors. The active corrosion protection mechanism is based on anion-exchange reactions induced by particular triggers, such as increases in the aggressiveness of the surrounding environment or initiation of the corrosion process [26,27].

The most common approach to synthesizing LDH is the in situ growth method, due to its low cost and ease of synthesizing at laboratory and industrial scales [28]. In this method, the source of Al^{3+} ions required to grow the LDH structure is the metallic substrate, and it results in coatings with better adherence compared to LDH coatings synthesized by other routes such as hydrothermal treatment [29–31], which is characterized by the use of autoclave to get high temperatures, or the urea hydrolysis method, which is used to form well-crystallized large LDH [32–34].

The knowledge of multifunctional surfaces with active protection based on ceramic-like coatings (PEO) is at an embryonic stage [35]. Y. Zhang [36] investigated the growth behavior of LDH layers on PEO-coated aluminum alloy. It was found that LDH grains are preferentially formed on the micro-pores of PEO coatings to provide effective film repairs. The most dominant factor which determines the PEO/LDH system behavior is the interaction between the electrolyte anions and $Al(OH)_2^+$ cations from PEO coating surface. This phenomenon has a major impact on the kinetic mechanism of the formation of LDH coating [37,38]. F. Chen [39] demonstrated that flakes were preferentially formed in the PEO pores at the initial stage of LDHs growth; it was also found that long-term corrosion resistance was significantly enhanced due to the limited penetration of corrosive ions [38]. The corrosive degradation of the substrate coated with PEO is accelerated by chloride ions in the environment. From a corrosion-relevance perspective, the entrapment of chloride anions via the ion-exchange mechanism is the most critical factor for designing efficient corrosion protection strategies. In fact, partial entrapment is possible when the PEO layer is sealed with LDH [40].

Although the concept of LDH-based post-treatment for sealing the PEO layers has been recently proven, many variables remain unexplored. In particular, this work is focused on the formation of in situ Zn-Al LDH on low energy consumption PEO coatings. For that, three PEO coatings were selected from 31 processes (with different combination of electrolytes, electrical conditions and time) based on corrosive behavior and energy consumption. The selected materials were thoroughly characterized and evaluated in terms of corrosion resistance.

2. Materials and Methods

2.1. Material

The composition of the 1050-H18 aluminum alloy (Famimetal S.L., Madrid, Spain) in wt.% is: 0.07 Zn, 0.05 Mn, 0.25 Si, 0.05 Cu, 0.40 Fe, 0.05 Mg, 0.05 Ti, 0.03 V and Al balance.

2.2. Specimens Preparation

The samples were cut from sheets into specimens of 30 × 20 × 1.5 mm^3 dimensions, ground to P1200 silicon carbide abrasive paper, rinsed in distilled water and methanol and dried in warm air. Prior to the PEO processing, specimens were etched in 15 wt.% sodium hydroxide solution for 20 s, rinsed in deionized water and dried in warm air. The working area was then limited to ~3 cm^2 using a commercial resin (Lacquer 45, MacDermid plc., Birmingham, UK).

2.3. Surface Treatment Based on PEO

PEO coatings were obtained using different electrolytes and conditions in order to grow thin PEO coatings with low energy consumption (Table 1) under vigorous agitation. The experimental system was carried out with a DC voltage/current-controlled power supply (SM400-AR-8 Systems electronic) equipped with a thermostatic jacket (25 ± 1 °C) under continuous electrolyte agitation. An AISI 316 steel plate of 7.5 × 15 cm^2 size was used as a counter electrode. After the PEO process, all samples were rinsed in distilled water and isopropanol and dried in warm air.

Table 1. PEO conditions of AA1050 alloy.

Coating	Electrolyte (g/L)		Coating	Electrolyte (g/L)		Coating	Electrolyte (g/L)	
	NaAlO$_2$	KOH		(Na$_3$P$_3$O$_6$)$_3$	KOH		Na$_2$SiO$_3$ **	KOH
A1.1	4	1.0	P1.1	30	1	S1.1	10.5	2.8
A1.2	4	3.3	P1.2	30	3	S1.2	10.5	3.5
A1.3	4	5.6	P1.3	30	5	S1.3	10.5	4.6
A2.1	7	1.0	P2.1	40	1	S2.1	15.0	2.8
A2.2	7	3.3	P2.2	40	3	S2.2	15.0	3.5
A2.3	7	5.6	P2.3	40	5	S2.3	15.0	4.6
A3.1	10	1.0	P3.1	50	1	S2.4	15.0	6.0
A3.2	10	3.3	P3.2	50	3	S2.5	15.0	8.0
A3.3	10	5.6	P3.3	50	5	S2.6	15.0	10.0
-	-	-	P4 *	Na$_4$P$_2$O$_7$: 20	2.8	S4 *	Na$_2$SiO$_3$·5H$_2$O: 25	2.8
-	-	-	P5 *	Na$_3$PO$_4$·12H$_2$O: 20	2.8	S5 *	Na$_2$SiO$_3$·5H$_2$O: 5	8.4

PEO treatment conditions: V (V): 400; j (A·cm^{-2}): 0.1; t (s): 180; PEO treatment conditions *: V (V): 350; j (A·cm^{-2}): 0.05; t (s): 200; ** Na$_2$SiO$_3$ alludes to the use of water glass with specific gravity of 1.3 g/L.

2.4. Synthesis of Zn-Al-LDH Growth

Zn-Al LDH-nitrate (LDH-NO$_3$) was synthesized on AA1050-H18 aluminium alloy. The specimens were immersed in the solution for different times under continuous stirring in order to form LDH (Table 2), then rinsed in deionized water and dried in air at room temperature.

Table 2. LDH synthesis conditions of AA1050-H18.

LDH Treatment	Chemical Composition (M)	Exposure Time (min)
1	Zn(NO$_3$)$_2$·6H$_2$O: 0.01 NH$_4$NO$_3$: 0.06	30
2	Zn(NO$_3$)$_2$·6H$_2$O: 0.01 NH$_4$NO$_3$: 0.06	60
3	Zn(NO$_3$)$_2$·6H$_2$O: 0.01 NaNO$_3$: 0.06	30
4	Zn(NO$_3$)$_2$·6H$_2$O: 0.01 NaNO$_3$: 0.06	60

All treatments were developed in 100 mL of aqueous solution at 95 °C. pH values were adjusted to 6.5 using 1 vol.% ammonia.

2.5. Characterization

Planar and cross-sectional views of the specimens were examined using a JEOL JSM 6335F (Tokyo, Japan) field emission scanning electron microscope (FESEM) working at 20 kV and equipped with an energy dispersive (EDS) spectrometer (OXFORD X-MAX, Oxford, UK). Coating cross-sections were ground through successive grades of SiC paper and polished to a 1 μm diamond finish.

Phase composition was examined by X-ray diffraction (XRD), with a Philips X'Pert MRD (Amsterdam, The Netherlands, Cu Kα = 1.54056 Å). The XRD patterns were taken using grazing incidence with a step size of (0.01°–1°) and a dwell time of 6 s per step at room temperature.

The specific energy consumption was calculated by integration of the voltage-time and current-time transients acquired by the power supply (SM 400AR-8 Systems electronic) during the PEO treatment (Equation (1)). With the obtained results, the energy consumption was calculated in terms of kW·h·m^{-2} according to Equation (1), and the specific energy consumption in kW·h·m^{-2}·µm^{-1} was obtained dividing P_{tot} by the coating thickness.

$$P_{tot} = \int_{t_0}^{t} [V \cdot j] \left[\frac{W \cdot s}{m^2}\right] \tag{1}$$

2.6. Electrochemical Behavior

Electrochemical impedance spectroscopy (EIS) was used to evaluate the corrosion resistance of the different coatings in an aqueous saline solution (NaCl 3.5 wt.%) at 25 °C. For that, a GillAC (ACM Instruments, Cumbria, UK) computer-controlled potentiostat and a three-electrode cell were used. The specimen was connected as a working electrode with an exposed area of 1 cm^2. A graphite electrode and a silver–silver chloride (Ag/AgCl) electrode used as the counter and the reference electrode, respectively. The solution inside the reference electrode was KCl 3 M, which provided a potential of 0.210 V with respect to the standard hydrogen electrode. The tests applying a sinusoidal perturbation of 10 mV RMS amplitude in the frequency range of 30 kHz–0.01 Hz were carried out after 1 h of immersion. All measurements were duplicated to ensure reproducibility.

3. Results and Discussion

3.1. PEO Coating Screening

The first screening process to select one PEO coating per electrolyte type was conducted in accordance with three factors: (i) presence of microdischarges during coating formation, (ii) visually uniform coating morphology and (iii) coating thickness ≥1 µm.

Among the different alternatives based on different electrolyte compositions, PEO coatings developed in phosphate and silicate electrolytes showed more promising results (with lower breakdown voltage values, beneficial for low energy consumption) (Table 3) and therefore additional process conditions were also tried (V (V): 350; j (A·cm^{-2}): 0.05; t (s): 200) (Table 1).

Table 3. Energy consumption values and electrolytes characteristic of selected flash PEO coatings.

Coating	Electrolyte Composition (g/L)	σ (mS/cm)	pH	U_{bd} (V)	Thickness (µm)	Growth Rate (µm·min^{-1})	Energy Consumption (KW·h·m^{-2}·µm^{-1})
A3.1	NaAlO$_2$: 10 KOH: 1	13.9	12.70	320	1.1 ± 0.3	0.37	4.98
P2.1	(Na$_3$P$_3$O$_6$)$_3$: 40 KOH: 1	11.1	12.5	257	2.4 ± 0.4	1.31	4.70
S4	Na$_2$SiO$_3$·5H$_2$O: 25 KOH: 2.8	32.9	12.7	108	1.3 ± 0.2	0.39	2.20

Finally, only the A3.1, A3.2, P2.1, S1.3, S2.6 and S4 PEO coatings (Table 1) fulfilled the previous requirements.

The last step of the screening process consisted of a corrosion evaluation based on the value of the modulus of the impedance obtained by electrochemical impedance spectroscopy (EIS). Figure 1 shows the |Z| × 10^{-2} Hz values of the studied materials, providing an estimation of the corrosion resistance, where higher values of |Z| indicate a lower corrosion rate [41].

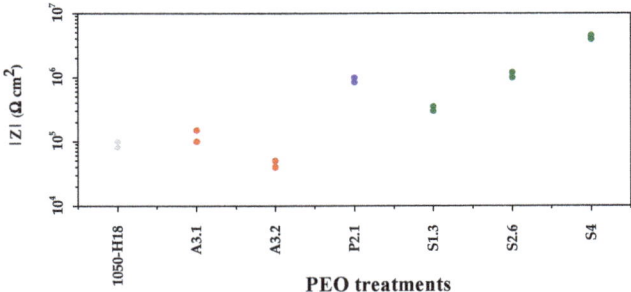

Figure 1. Scatter diagram of impedance modulus at 10^{-2} Hz; selected PEO coatings on AA1050 alloy, citing two measurements per condition.

With the aim of studying the influence of electrolytes on LDH growth, three PEO coatings were selected (one per electrolyte composition: aluminate A3.1, phosphate 2.1 and silicate S4). The specific energy consumption of selected coatings was calculated (Table 3) by integration of the voltage-time and current-time transients (Figure 2) recorded during the PEO process to verify that the developed coatings were energy efficient. The obtained values reveal the effect of electrolyte composition on energy consumption, coating growth rate and breakdown voltage values (Table 3).

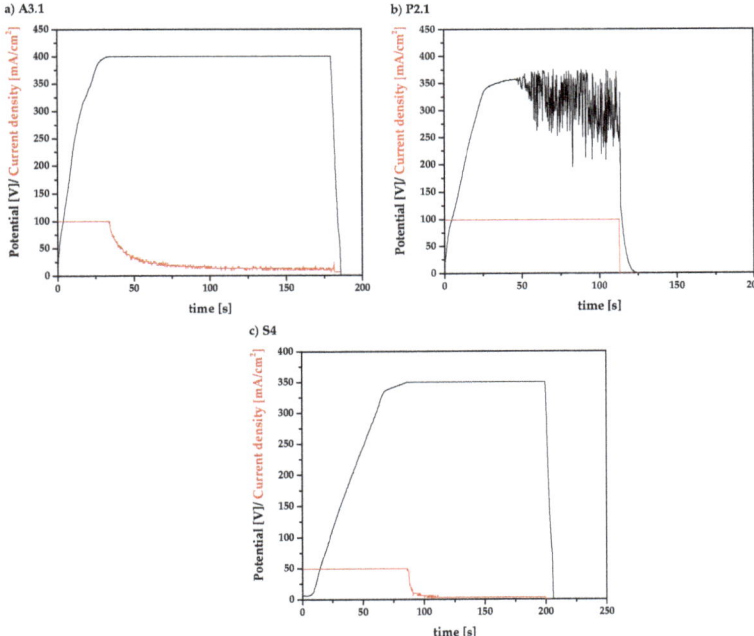

Figure 2. Voltage-current time curves for (**a**) A3.1, (**b**) P2.1 and (**c**) S4 PEO coatings on AA1050 alloy.

As can be seen in Figure 2, the current drop was observed only in aluminate and silicate electrolyte cases (Figure 2a,c) when 400 and 350 V limitations (Table 1) were achieved after 40 and 75 s of treatment, respectively, and the power supply switched to a constant voltage-control mode. Typically, when the current density was below 20 mA·cm^{-2}, the microdischarges extinguished; the anodizing, however, was carried on till set time in order to repair microdefects in the oxide material [42].

The fact that the limiting voltage was reached (hence the current drop) could be explained by high coating density and, consequently, higher resistance of the oxide to charge and mass transfer that were achieved at early stages [18]. High value of energy consumption in the case of aluminate is mainly due to the fact that aluminate species in the electrolyte gave rise to the formation of coating composed of nearly pure alumina, which has very low electron conductivity (i.e., the current flows mainly by ion and not by electron transfer) [12]. Further, the high breakdown voltage (320 V) in the aluminate electrolyte compelled the use of a higher voltage limit (400 V) in order to ensure a long enough period of sparking in order to achieve a uniform coating of a significant thickness; this yielded a higher specific energy consumption value.

As a result, the dielectric breakdown voltage was high, the limiting set voltage was achieved quickly and, as a consequence of the current drop, the sparking period was short, hence the low coating growth rate [22,43]. Similarly, high value of energy consumption in the case of phosphate electrolyte, where sparking was observed until the end of the treatment and coating growth rate was relatively high, was due to the absence of current drop, because the limiting 350 V were never reached. In this case, higher resistance of the oxide to charge and mass transfer were achieved at 60 s, giving rise to intense microdischarges and voltage variations, and therefore the treatment was stopped at 115 s in order to maintain coating uniformity.

The lowest energy consumption of 2.2 kW·h·m^{-2}·μm^{-1} was achieved in case of silicate electrolyte. This was mainly the result of its high electrical conductivity and, therefore, low U_{bd}. The onset of microdischarges early in the treatment and the relatively long sparking period before current decay resulted in the intermediate coating growth rate value (Table 3).

The specific energy consumption values obtained under DC conditions in the present work are similar to those reported in studies carried out under AC conditions. For example, E. Matykina et al. developed a PEO coating on pure aluminium using silicate electrolyte, and obtained a growth rate of 1.3 μm·min^{-1} and energy consumption values of 4.77 KW·h·m^{-2}·μm^{-1} [18]. Y.L. Cheng et al. reported a growth rate of 11.3 μm·min^{-1} and energy consumption values of 5 KW·h·m^{-2}·μm^{-1} for Al-Cu alloy in concentrated aluminate electrolyte [44]. It is well known that DC conditions promote low growth rates in comparison with AC conditions [18,22]; however, in this study the values obtained were considerably lower compared with the available data for different PEO treatments on commercial Al alloys, which can be as high as 26.7 kW·h·m^{-2}·μm^{-1} [18]. The present findings demonstrate that in order to reduce specific energy consumption under DC conditions it is necessary to (i) limit the final forming voltage that ensures a current drop, and (ii) use electrolytes with conductivity, which ensures low microdischarges onset voltages and extended sparking periods, as in the case of the S4 electrolyte.

3.2. PEO Coatings Characterization

Figure 3 shows the planar view and cross-section scanning electron micrographs of AA1050 coated by selected PEO coatings. All selected treatments show a thin oxide layer of 1–2.5 μm (Table 3). This is particularly evident in aluminate electrolyte-based PEO coating (A3.1), where the Al-Fe intermetallic compounds from the substrate are still visible in the coating (Figure 3a, inset) due to its low thickness. This coating is also more heterogeneous (Figure 3b) than the rest, which is attributable to its high breakdown voltage values (Figure 2) [37,45,46] and, as mentioned before, its low coating growth rate. Phosphate electrolyte-based PEO coating (P2.1) showed a homogeneous (Figure 3c) surface appearance (Figure 3c), and the highest thickness value (Figure 3d). This was mainly due to the presence of polyphosphate species that participated in PEO coating formation and favored its high coating growth rate [47]. Silicate electrolyte-based PEO coating (S4) (Figure 3e), with the lowest breakdown voltage, showed a homogeneous surface morphology with very sparse submicrometric pores. The latter may be attributable to the formation of a thin superficial glassy layer of SiO_2, which can be surmised from the EDS analysis where the presence of 1.5 at.% Si in the coatings and a greater content of oxygen than in the other two coatings was confirmed.

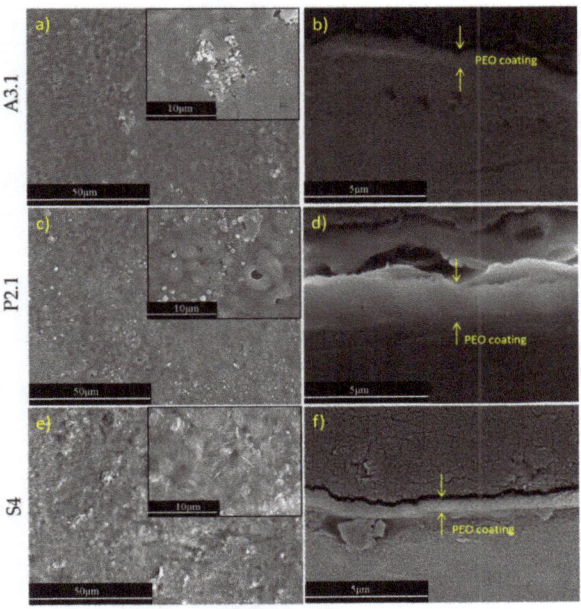

PEO Coating	EDS (at.%)				
	O	Al	Fe	P	Si
A3.1	45.7	52.8	1.1	–	–
P2.1	53.3	46.6	–	1.1	–
S4	64.3	33.6	–	–	1.5

Figure 3. Planar view (**a,c,e**) and cross-section view (**b,d,f**) of secondary electron images of the PEO coatings (A3.1, P2.1, S4), respectively. EDS analysis was performed on the areas corresponding to the planar views of the coatings.

3.3. LDH Screening

The effect of reactant composition and treatment time during the growth of LDH coatings were investigated. Figure 4 depicts the XRD patterns of the different LDH treatments (Table 2) grown on the bulk material. The presence of peaks at 9.6° and 19.9° corresponding to the characteristic (003) and (006) reflections of LDHs intercalated with NO^{3-} [32,48,49] indicates the formation of LDH under the different conditions.

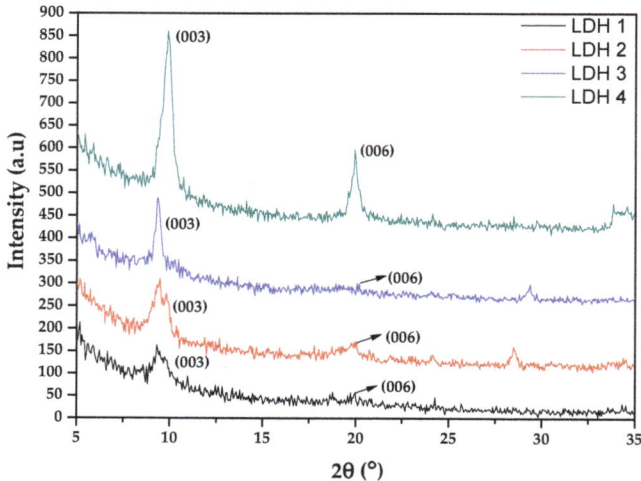

Figure 4. XRD patterns of Zn-Al-LDH-coated AA1050 alloy at different conditions.

It was revealed that treatments containing NaNO$_3$ in the solution led to more defined and intense peaks, probably due to the presence of sodium ions in the LDH gallery [50]. On the contrary, the presence of NH$_4$NO$_3$ drove the formation of broadened peaks [51]. In fact, just in the case of LDH grown in NH$_4$NO$_3$, a small peak was revealed at 9.9° that could be associated with an LDH phase intercalated with carbonate, due to the formation of LDH layers under atmospheric conditions [52].

LDHs formed under long treatment times (LDH 2 and 4) showed very strong peaks in comparison with LDHs formed under short treatment times (LDH 1 and 3), mainly because an increment in the LDH degree of crystallinity took place [51].

The correlation between XRD patterns of studied LDH coatings and planar view scanning electron micrographs were investigated. Figure 5 shows secondary electron images of the LDH treatments (Table 2) grown on the pure aluminium.

The typical flake-like LDH structure could be clearly observed for LDH carried out in NH$_4$NO$_3$ at short treatment times, whereas LDH developed in NaNO$_3$ formed this structure at long exposure times. According to Figure 5a, the LDH structure carried out in the presence of NH$_4$NO$_3$ is in good agreement with the XRD patterns that showed broadened peaks and, consequently, a highly open LDH structure. Additionally, Figure 5b clearly shows (also consistent with the XRD pattern) a non-defined LDH structure, which is usually attributed to the incorporation of carbonate ions into the LDH gallery [48]. However, LDH developed in NaNO$_3$ showed typical curved plate-like LDH microcrystals at long exposure times, and a non-defined structure at short exposure times (Figure 5c,d). This was mainly due to an increment in the degree of LDH crystallinity that was observed in XRD patterns (Figure 4).

In order to evaluate the correlation between the corrosion protection and the structure of studied LDHs, a screening process based on corrosion performance (EIS) was carried out. Figure 6 depicts the Bode and Nyquist plots for AA1050 alloy with studied LDH coatings.

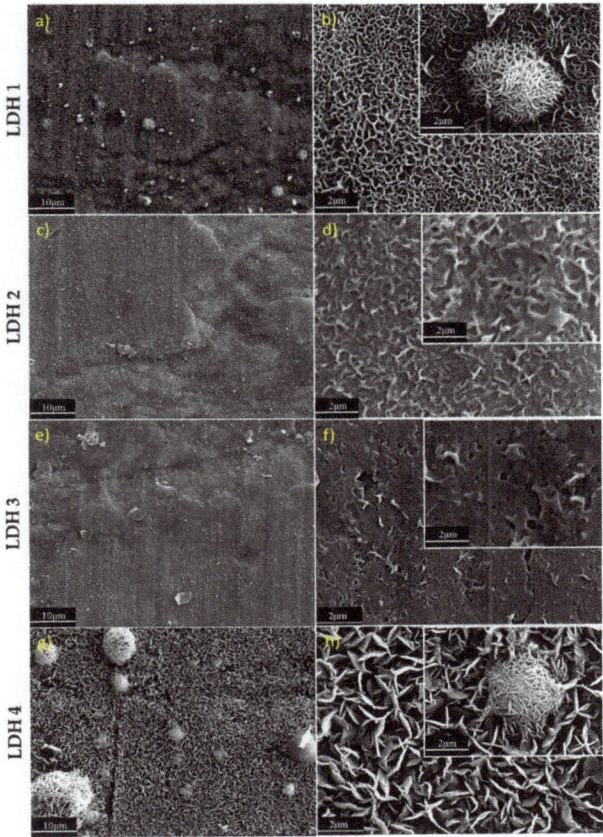

Figure 5. Secondary electron images of the LDH 1 (**a,b**) LDH 2 (**c,d**), LDH 3 (**e,f**) and LDH 4 (**g,h**) coatings.

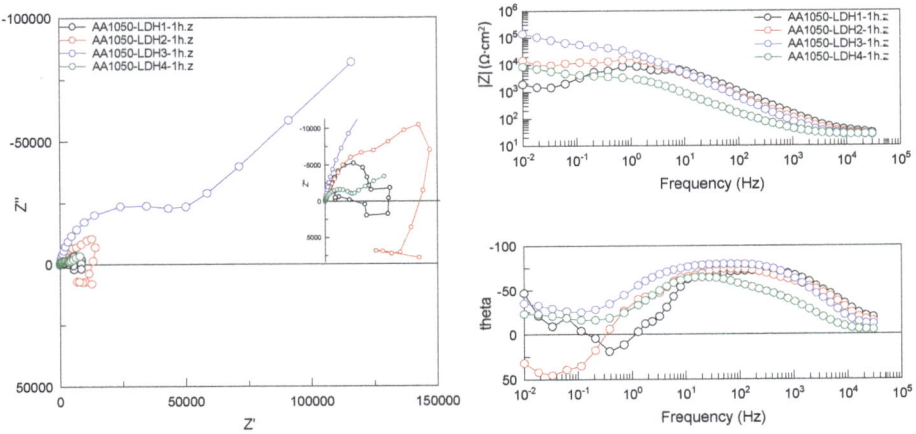

Figure 6. Bode plots for AA1050 alloy with studied LDH coatings.

119

From the point of view of coating structure, the presence of spheroidal particles in flake-like LDH 1 and LDH 4 (Figure 5) [14] was associated with the presence of secondary phases [35–37] that favored aluminium cation dissolution due to their highly cathodic behavior [48]. In addition, the porous structure of these spheroidal particles also favored Cl$^-$ anion penetration into the LDH gallery, and for this reason these coatings showed the lowest corrosion protection among the studied LDH coatings (Figure 6) [53–55]. From the point of view of corrosion protection of non-defined LDH coatings, it should be noted that LDH 2 showed similar corrosion behavior in comparison with LDH 4, which can be attributed to its intermediate non-porous LDH structure (Figure 5). On the contrary, LDH 3 provided a beneficial effect to corrosion protection in comparison with all the studied LDH coatings. This may be due to the presence of sodium ions in the LDH gallery, which facilitate the formation of non-porous corrosion-protective LDH coating (Figure 6). For this reason, LDH 3 was the selected treatment to use for the selected PEO coatings and study their anti-corrosion properties (Figure 6).

3.4. PEO-LDH Coating Characterization

Figure 7 depicts the XRD patterns of selected PEO coatings (A3.1, P2.1 and S4) with LDH 3 and discloses the effects of the PEO coating compositions on LDH growth.

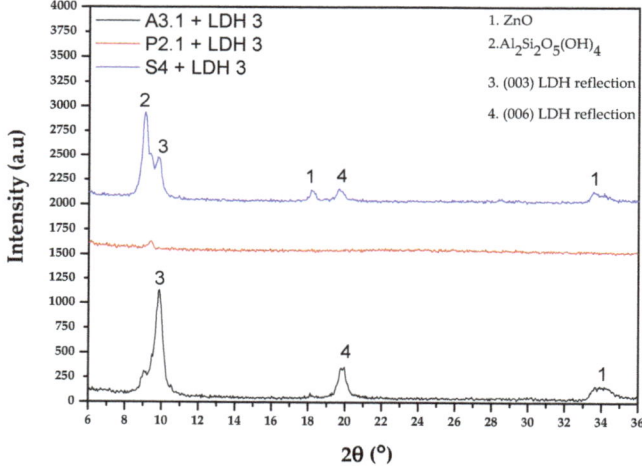

Figure 7. XRD patterns of Zn-Al-LDH-coated A3.1, P2.1 and S4 PEO coatings.

PEO coatings based on aluminate and silicate electrolytes showed the typical peaks at 9.6° and 19.9°, which corresponded to the characteristic (003) and (006) reflections of LDH intercalated with NO$_3^-$ [32,48,49]. Additionally, the presence of ZnO and Al$_2$Si$_2$O$_5$(OH)$_4$ characteristic peaks in the S4-LDH XRD pattern came from LDH chemical composition (Table 2) and silicate electrolytes, respectively (Table 1).

In the particular case of PEO coating developed in phosphate electrolyte, there were no peaks detected in that range. This could be attributed to the formation of non-crystalline phases, or to only a small amount that could not be detected at the selected scan rate (Figure 7).

Figure 8 highlights a detailed morphology of the selected PEO-LDH coatings, which reveals the importance of PEO coating composition. The characteristic flake-like LDH structure can be clearly observed for the A3.1-LDH coating, in which LDH flakes are covering the whole PEO coating (Figure 8a,b). This is also observable in the S4-LDH coating, but in this case is more heterogenous (Figure 8e,f). These results are in accordance with XRD patterns that showed the presence of these characteristic reflections (Figure 7). However, for the P2.1-LDH coating there is a drastic decrease of the density of LDH-like flakes (Figure 8c,d), which is in accordance with the XRD patterns (Figure 7).

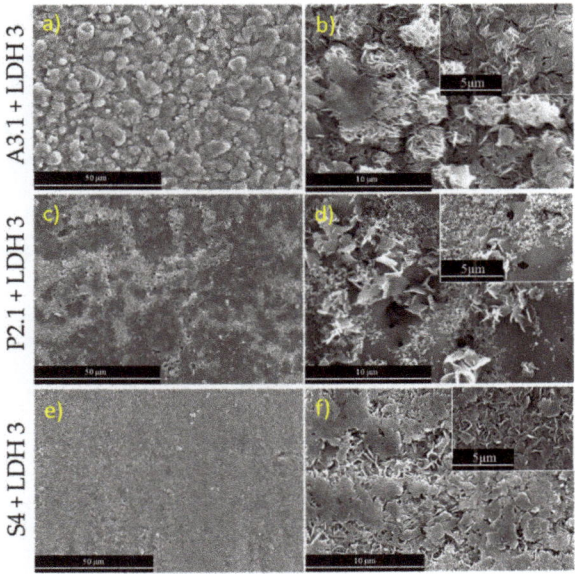

Figure 8. Secondary electron images of the planar view of the PEO coatings after LDH 3 coating formation: (**a,b**) A3.1, (**c,d**) P2.1, (**e,f**) S4.

This fact is in strong agreement with a high dependence on the availability of $Al(OH)_2^+$ cations necessary to form Zn-Al-LDH. According to previous studies [40,56], Zn-Al LDH synthesis can be explained via the following chemical reactions:

$$Al_2O_3 + 3H_2O \rightarrow 2Al(OH)_3 \tag{2}$$

$$Al(OH)_3 + NH_4^+ \rightarrow Al(OH)_2^+ + NH_3 \cdot H_2O \tag{3}$$

$$Zn(OH) + Al(OH)_2^+ + 2NO_3 \rightarrow LDH - NO_3 \tag{4}$$

As mentioned before, the in situ growth method was used in the present work and, consequently, $Al(OH)_2^+$ cations were an essential requirement to form the LDH layers. Due to the porous structure of PEO coatings and their compositions, two sources can provide $Al(OH)_2^+$ cations: (i) the aluminium metal matrix (due to the electrochemical interactions with LDH solutions), and PEO coating thickness [57].

Firstly, in consideration of the PEO coating cross-section (Figure 3) and planar view micrographs after LDH treatment (Figure 8), it can be concluded that the amount of LDH flakes on the selected PEO coatings was highest for A3.1, and lowest for the P2.1 and S4 coatings. This could firstly be explained by thickness, and secondly by the chemical composition of the PEO coating surfaces. Due to the low thickness of the A3.1 PEO coating (~1 μm) (Figure 3a,b), the migration capacity of $Al(OH)_2^+$ cations from the aluminium metal matrix towards the coating surface was ensured. However, the P2.1 (Figure 3c,d) and S4 (Figure 3e,f) PEO coatings showed higher thickness values (~1.5–2 μm), and, consequently, the migration capacity of $Al(OH)_2^+$ was reduced. With respect to the chemical composition of PEO coating surfaces, the aluminium content decreased in the order A3.1 > P2.1 > S4 (Figure 3, EDS analysis table) due to the presence of aluminate species in the A3.1 electrolyte. Further, greater charge passed during the P2.1 treatment, and intense sparking from the onset of the microdischarges until the end of the treatment (Figure 2b) resulted in greater thickness and density of the P2.1 coatings. Consequently, a highly heterogeneous LDH layer was achieved in comparison with the S4 PEO coating. This justifies the absence of peaks in XRD patterns of the P2.1-LDH (Figure 7).

3.5. Corrosion Resistance of PEO + LDH Coatings

In order to evaluate the effect of LDH formation on selected PEO coatings, corrosion resistance was measured by electrochemical impedance spectroscopy (EIS) for 1 h of immersion in 3.5 wt.% NaCl solution at room temperature (Figure 9).

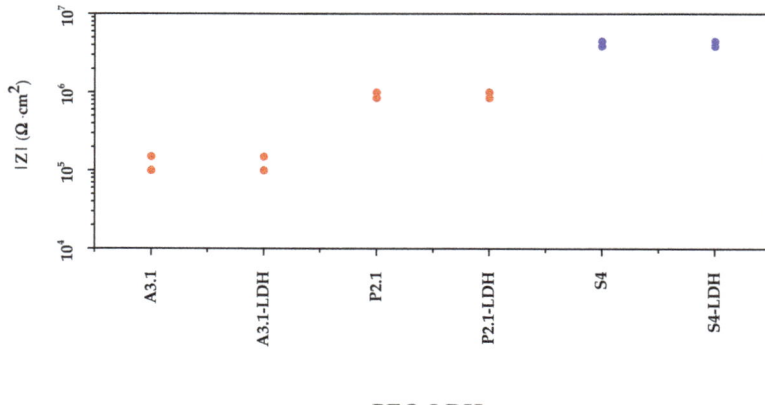

Figure 9. Scatter diagram of impedance modulus at 10^{-2} Hz; selected PEO-LDH coatings on AA1050 alloy.

As can be seen in Figure 9, PEO coatings with and without LDH treatment showed high similitude of the impedance modulus.

It is important to note that, in this work, the corrosion resistance of different PEO coatings was strongly connected with PEO coating porosity, because the lack of pores restricts the Cl$^-$ penetration and prevents its detrimental effects [58]. For this reason, A3.1 showed the lowest corrosion resistance due to the combination of pores and heterogeneities in comparison with the P2.1 and S4 coatings. The highest corrosion protection exhibited by the S4 PEO coating (~5 × 10^6 Ω·cm^2) may be attributed to its sparse surface porosity due to the formation of a glassy silica-rich layer.

In this work, the results showed similar behavior of PEO-LDH compared with PEO coatings without any clear improvement after the post-treatment. This could be attributed to several factors, for instance: (i) LDH flake resistance was negligible compared with that of the PEO coating, and (ii) PEO coating lost some of its barrier properties during the formation of LDH, somewhat retracting from the possible small beneficial effect of the LDH layer.

It should be noted that no studies of LDH formation on flash PEO coatings (at <5 min anodizing time) have been reported so far. However, based on this work and few works carried out with non-flash PEO of aluminium, some conclusions can be drawn regarding the effects of LDH post-treatments. For instance, when an LDH layer is not loaded with inhibitors, the corrosion resistance remains unchanged or degrades slightly, as has been shown in [37]. On the other hand, when LDH is intercalated with an inhibitor (e.g., vanadate ions), an improvement in corrosion resistance is observed with immersion time due to an active protection effect [37,38].

The present findings highlight that the development of LDH-container layers did not deteriorate the corrosion resistance of flash PEO coatings, which has a potential for added active protection functionality. Therefore, the first stage of active protection system building can be considered successful. The second stage would consist of loading the LDH scaffold with corrosion inhibitors that would ensure enhanced corrosion protection.

In conclusion, these results are highly relevant for understanding the relation between coherent and uniform LDH layer formations on flash PEO coatings, which is the first step in achieving active protection systems through the incorporation of green corrosion inhibitors into the LDH layer.

4. Conclusions

The following can be summarized from this preliminary study of LDH growth on flash PEO coatings:

- Flash PEO coatings with ~1–2 µm thickness and ~2–5 KW·h·m^{-2}·µm^{-1} energy consumption were genereated on a commercially pure aluminum alloy. Low energy consumption was ensured through relatively high electrolyte conductivity and a transition of the anodizing regime from constant current to constant voltage control.
- The first stage of the active protection system was successfully completed on flash PEO coatings via the development of an LDH layer. LDH coating is continuous and well defined when the PEO layer is thin (~1 µm), and the LDH formation is further facilitated when additional Al(OH)$_2^+$ cations are lixiviated from the coating.
- Corrosion resistance of inhibitor-free flash PEO/LDH coatings is mainly determined by the low porosity of the PEO layer. Formation of the LDH layer does not compromise the corrosion resistance of flash PEO coatings. Loading of the LDH scaffold with corrosion inhibitors is necessary in order to achieve an enhanced corrosion protection.

Author Contributions: Formal Analysis, R.d.O., M.M., B.M., R.A. and E.M.; Funding Acquisition, M.M.; Investigation, R.d.O. and M.M.; Methodology, R.d.O. and M.M.; Resources, M.M. and E.M.; Supervision, R.A. and E.M.; Writing—Original Draft, R.d.O., M.M. and B.M.; Writing—Review & Editing, R.A. and E.M.; Conceptualization, R.d.O., M.M., B.M., R.A. and E.M.; Software, R.d.O., M.M., B.M., R.A. and E.M.; Validation, R.d.O, M.M., B.M., R.A. and E.M.; Data Curation, R.d.O., M.M. and B.M.; Visualization, R.d.O., M.M., B.M., R.A. and E.M.; Project Administration R.A. and E.M.

Funding: This work was partially supported by (MAT2015-73355-JIN), ADITIMAT-CM (S2018/NMT-4411) and RTI2018-096391-B-C33 MCIU/AEI/FEDER, UE. MM is grateful to the Ramon y Cajal Programme (MICINN, Spain, RYC-2017-21843).

Conflicts of Interest: The authors declare no conflict of interest.

References

1. Matykina, E.; Arrabal, R.; Skeldon, P.; Thompson, G. Investigation of the growth processes of coatings formed by AC plasma electrolytic oxidation of aluminium. *Electrochim. Acta* **2009**, *54*, 6767–6778. [CrossRef]
2. Arrabal, R.; Matykina, E.; Hashimoto, T.; Skeldon, P.; Thompson, G. Characterization of AC PEO coatings on magnesium alloys. *Surf. Coat. Technol.* **2009**, *203*, 2207–2220. [CrossRef]
3. Matykina, E.; Skeldon, P.; Thompson, G.E. Fundamental and practical evaluations of PEO coatings of titanium. *Int. Heat Surf. Eng.* **2009**, *3*, 45–51. [CrossRef]
4. Abrahami, S.T.; De Kok, J.M.M.; Mol, J.M.C.; Terryn, H. Towards Cr(VI)-free anodization of aluminum alloys for aerospace adhesive bonding applications: A review. *Front. Chem. Sci. Eng.* **2017**, *11*, 465–482. [CrossRef]
5. Kulinich, S.; Akhtar, A.S. On conversion coating treatments to replace chromating for Al alloys: Recent developments and possible future directions. *Russ. J. Non-Ferrous Met.* **2012**, *53*, 176–203. [CrossRef]
6. Gharbi, O.; Thomas, S.; Smith, C.; Birbilis, N. Chromate replacement: What does the future hold? *npj Mater. Degrad.* **2018**, *2*, 12. [CrossRef]
7. Khan, R.; Yerokhin, A.; Li, X.; Dong, H.; Matthews, A. Surface characterisation of DC plasma electrolytic oxidation treated 6082 aluminium alloy: Effect of current density and electrolyte concentration. *Surf. Coat. Technol.* **2010**, *205*, 1679–1688. [CrossRef]
8. Guan, Y.; Xia, Y.; Li, G. Growth mechanism and corrosion behavior of ceramic coatings on aluminum produced by autocontrol AC pulse PEO. *Surf. Coat. Technol.* **2008**, *202*, 4602–4612. [CrossRef]

9. Han, I.; Choi, J.H.; Zhao, B.H.; Baik, H.K.; Lee, I.-S. Changes in anodized titanium surface morphology by virtue of different unipolar DC pulse waveform. *Surf. Coat. Technol.* **2007**, *201*, 5533–5536. [CrossRef]
10. Yerokhin, A.; Shatrov, A.; Samsonov, V.; Shashkov, P.; Pilkington, A.; Leyland, A.; Matthews, A. Oxide ceramic coatings on aluminium alloys produced by a pulsed bipolar plasma electrolytic oxidation process. *Surf. Coat. Technol.* **2005**, *199*, 150–157. [CrossRef]
11. Dunleavy, C.; Curran, J.; Clyne, T. Time dependent statistics of plasma discharge parameters during bulk AC plasma electrolytic oxidation of aluminium. *Appl. Surf. Sci.* **2013**, *268*, 397–409. [CrossRef]
12. Matykina, E.; Arrabal, R.; Skeldon, P.; Thompson, G.; Belenguer, P. AC PEO of aluminium with porous alumina precursor films. *Surf. Coat. Technol.* **2010**, *205*, 1668–1678. [CrossRef]
13. Curran, J.; Clyne, T. Thermo-physical properties of plasma electrolytic oxide coatings on aluminium. *Surf. Coat. Technol.* **2005**, *199*, 168–176. [CrossRef]
14. Barik, R.; Wharton, J.; Wood, R.; Stokes, K.; Jones, R.; Wharton, J. Corrosion, erosion and erosion–corrosion performance of plasma electrolytic oxidation (PEO) deposited Al_2O_3 coatings. *Surf. Coat. Technol.* **2005**, *199*, 158–167. [CrossRef]
15. Srinivasan, P.B.; Liang, J.; Blawert, C.; Stormer, M.; Dietzel, W. Effect of current density on the microstructure and corrosion behaviour of plasma electrolytic oxidation treated AM50 magnesium alloy. *Appl. Surf. Sci.* **2009**, *255*, 4212–4218. [CrossRef]
16. Aliasghari, S. Plasma Electrolytic Oxidation of Titanium. Ph.D. Thesis, The University of Manchester, Manchester, UK, September 2014.
17. Mohedano, M.; Lu, X.; Matykina, E.; Blawert, C.; Arrabal, R.; Zheludkevich, M.L. Plasma electrolytic oxidation (PEO) of metals and alloys. In *Encyclopedia of Interfacial Chemistry*, 1st ed.; Wandelt, K., Ed.; Elsevier: Amsterdam, The Netherlands, 2018; pp. 423–438.
18. Matykina, E.; Arrabal, R.; Pardo, A.; Mohedano, M.; Mingo, B.; Rodríguez, I.; González, J. Energy-efficient PEO process of aluminium alloys. *Mater. Lett.* **2014**, *127*, 13–16. [CrossRef]
19. Sinko, J. Challenges of chromate inhibitor pigments replacement in organic coatings. *Prog. Org. Coat.* **2001**, *42*, 267–282. [CrossRef]
20. Snizhko, L.; Yerokhin, A.; Gurevina, N.; Patalakha, V.; Matthews, A.; Snizhko, L. Excessive oxygen evolution during plasma electrolytic oxidation of aluminium. *Thin Solid Films* **2007**, *516*, 460–464. [CrossRef]
21. Snizhko, L.; Yerokhin, A.; Pilkington, A.; Gurevina, N.; Misnyankin, D.; Leyland, A.; Matthews, A.; Snizhko, L. Anodic processes in plasma electrolytic oxidation of aluminium in alkaline solutions. *Electrochim. Acta* **2004**, *49*, 2085–2095. [CrossRef]
22. Matykina, E.; Arrabal, R.; Mohedano, M.; Mingo, B.; Gonzalez, J.; Pardo, A.; Merino, M. Recent advances in energy efficient PEO processing of aluminium alloys. *Trans. Nonferrous Met. Soc. China* **2017**, *27*, 1439–1454. [CrossRef]
23. Yasakau, K.; Tedim, J.; Zheludkevich, M.; Ferreira, M.; Yasakau, K.; Zheludkevich, M. Smart self-healing coatings for corrosion protection of aluminium alloys. In *Handbook of Smart Coatings for Materials Protection*, 1st ed.; Woodhead Publishing: Cambridge, UK, 2014; pp. 224–274.
24. Mardel, J.; Garcia, S.; Corrigan, P.; Markley, T.; Hughes, A.; Muster, T.; Lau, D.; Harvey, T.; Glenn, A.; White, P.; et al. The characterisation and performance of $Ce(dbp)_3$-inhibited epoxy coatings. *Prog. Org. Coat.* **2011**, *70*, 91–101. [CrossRef]
25. Osborne, J.H.; Blohowiak, K.Y.; Taylor, S.; Hunter, C.; Bierwagon, G.; Carlson, B.; Bernard, D.; Donley, M.S. Testing and evaluation of nonchromated coating systems for aerospace applications. *Prog. Org. Coat.* **2001**, *41*, 217–225. [CrossRef]
26. Guo, L.; Wu, W.; Zhou, Y.; Zhang, F.; Zeng, R.; Zeng, J. Layered double hydroxide coatings on magnesium alloys: A review. *J. Mater. Sci. Technol.* **2018**, *34*, 1455–1466. [CrossRef]
27. Williams, G.; Geary, S.; McMurray, H.; McMurray, H. Smart release corrosion inhibitor pigments based on organic ion-exchange resins. *Corros. Sci.* **2012**, *57*, 139–147. [CrossRef]
28. Cavani, F.; Trifirò, F.; Vaccari, A. Hydrotalcite-type anionic clays: Preparation, properties and applications. *Catal. Today* **1991**, *11*, 173–301. [CrossRef]
29. Guo, X.; Xu, S.; Zhao, L.; Lu, W.; Zhang, F.; Evans, D.G.; Duan, X. One-step hydrothermal crystallization of a layered double hydroxide/alumina bilayer film on aluminum and its corrosion resistance properties. *Langmuir* **2009**, *25*, 9894–9897. [CrossRef] [PubMed]

30. Zhang, F.; Zhang, C.-L.; Song, L.; Zeng, R.-C.; Liu, Z.-G.; Cui, H.-Z. Corrosion of in-situ grown MgAl-LDH coating on aluminum alloy. *Trans. Nonferrous Met. Soc. China* **2015**, *25*, 3498–3504. [CrossRef]
31. Xu, Z.P.; Lu, G.Q. Hydrothermal synthesis of layered double hydroxides (LDHs) from mixed MgO and Al_2O_3: LDH formation mechanism. *Chem. Mater.* **2005**, *17*, 1055–1062. [CrossRef]
32. Hao, L.; Yan, T.; Zhang, Y.; Zhao, X.; Lei, X.; Xu, S.; Zhang, F. Fabrication and anticorrosion properties of composite films of silica/layered double hydroxide. *Surf. Coat. Technol.* **2017**, *326*, 200–206. [CrossRef]
33. Zhang, Y.; Li, Y.; Ren, Y.; Wang, H.; Chen, F. Double-doped LDH films on aluminum alloys for active protection. *Mater. Lett.* **2017**, *192*, 33–35. [CrossRef]
34. Staal, L.B.; Pushparaj, S.S.C.; Forano, C.; Prevot, V.; Ravnsbæk, D.B.; Bjerring, M.; Nielsen, U.G. Competitive reactions during synthesis of zinc aluminum layered double hydroxides by thermal hydrolysis of urea. *J. Mater. Chem. A* **2017**, *5*, 21795–21806. [CrossRef]
35. Dou, B.; Wang, Y.; Zhang, T.; Liu, B.; Shao, Y.; Meng, G.; Wang, F. Growth behaviors of layered double hydroxide on microarc oxidation film and anti-corrosion performances of the composite film. *J. Electrochem. Soc.* **2016**, *163*, C917–C927. [CrossRef]
36. Zhang, Y. Investigating the growth behavior of LDH layers on MAO-coated aluminum alloy: Influence of microstructure and surface element. *Int. J. Electrochem. Sci.* **2018**, *13*, 610–620. [CrossRef]
37. Mohedano, M.; Serdechnova, M.; Starykevich, M.; Karpushenkov, S.; Bouali, A.; Ferreira, M.; Zheludkevich, M. Active protective PEO coatings on AA2024: Role of voltage on in-situ LDH growth. *Mater. Des.* **2017**, *120*, 36–46. [CrossRef]
38. Serdechnova, M.; Mohedano, M.; Kuznetsov, B.; Mendis, C.L.; Starykevich, M.; Karpushenkov, S.; Tedim, J.; Ferreira, M.G.S.; Blawert, C.; Zheludkevich, M.L. PEO coatings with active protection based on in-situ formed LDH-nanocontainers. *J. Electrochem. Soc.* **2017**, *164*, C36–C45. [CrossRef]
39. Chen, F.; Yu, P.; Zhang, Y. Healing effects of LDHs nanoplatelets on MAO ceramic layer of aluminum alloy. *J. Alloy. Compd.* **2017**, *711*, 342–348. [CrossRef]
40. Tedim, J.; Kuznetsova, A.; Salak, A.; Montemor, F.; Snihirova, D.; Pilz, M.; Zheludkevich, M.; Ferreira, M.; Salak, A.; Zheludkevich, M. Zn–Al layered double hydroxides as chloride nanotraps in active protective coatings. *Corros. Sci.* **2012**, *55*, 1–4. [CrossRef]
41. Liu, Y.; Yin, X.; Zhang, J.; Yu, S.; Han, Z.; Ren, L. A electro-deposition process for fabrication of biomimetic super-hydrophobic surface and its corrosion resistance on magnesium alloy. *Electrochim. Acta* **2014**, *125*, 395–403. [CrossRef]
42. Tsunekawa, S.; Aoki, Y.; Habazaki, H. Two-step plasma electrolytic oxidation of Ti–15V–3Al–3Cr–3Sn for wear-resistant and adhesive coating. *Surf. Coat. Technol.* **2011**, *205*, 4732–4740. [CrossRef]
43. Baron-Wiechec, A.; Burke, M.; Hashimoto, T.; Liu, H.; Skeldon, P.; Thompson, G.; Habazaki, H.; Ganem, J.-J.; Vickridge, I. Tracer study of pore initiation in anodic alumina formed in phosphoric acid. *Electrochim. Acta* **2013**, *113*, 302–312. [CrossRef]
44. Cheng, Y.; Cao, J.; Mao, M.; Xie, H.; Skeldon, P. Key factors determining the development of two morphologies of plasma electrolytic coatings on an Al–Cu–Li alloy in aluminate electrolytes. *Surf. Coat. Technol.* **2016**, *291*, 239–249. [CrossRef]
45. Sykes, J.; Thompson, G.E.; Mayo, D.; Skeldon, P. Anodic film formation on high strength aluminium alloy FVS0812. *J. Mater. Sci.* **1997**, *32*, 4909–4916. [CrossRef]
46. Fratila-Apachitei, L.; Tichelaar, F.; Thompson, G.; Terryn, H.; Skeldon, P.; Duszczyk, J.; Katgerman, L. A transmission electron microscopy study of hard anodic oxide layers on AlSi(Cu) alloys. *Electrochim. Acta* **2004**, *49*, 3169–3177. [CrossRef]
47. Guo-Hua, L.; Wei-Chao, G.; Huan, C.; Li, L.; Er-Wu, N.; Si-Ze, Y. Microstructure and corrosion performance of oxide coatings on aluminium by plasma electrolytic oxidation in silicate and phosphate electrolytes. *Chin. Phys. Lett.* **2006**, *23*, 3331–3333. [CrossRef]
48. Cao, Y.; Zheng, D.; Li, X.; Lin, J.; Wang, C.; Dong, S.; Lin, C. Enhanced corrosion resistance of superhydrophobic layered double hydroxide (LDH) films with long-term stability on Al substrate. *ACS Appl. Mater. Interfaces* **2018**, *10*, 15150–15162. [CrossRef]
49. Zhang, M.; Ma, L.; Sun, Y.; Liu, Y.; Wang, L.-L. Insights into the use of metal-organic framework as high performance anti-corrosion coatings. *ACS Appl. Mater. Interfaces* **2018**, *10*, 2259–2263. [CrossRef]

50. Ay, A.N.; Mafra, L.; Zümreoglu-Karan, B.; Zümreoglu-Karan, B.; Zümreoglu-Karan, B. A simple mechanochemical route to layered double hydroxides: Synthesis of hydrotalcite-like Mg-Al-NO$_3$-LDH by manual grinding in a mortar. *Zeitschrift für anorganische und allgemeine Chemie* **2009**, *635*, 1470–1475. [CrossRef]
51. Sertsova, A.; Subcheva, E.N.; Yurtov, E. Synthesis and study of structure formation of layered double hydroxides based on Mg, Zn, Cu, and Al. *Russ. J. Inorg. Chem.* **2015**, *60*, 23–32. [CrossRef]
52. Tedim, J.; Zheludkevich, M.; Bastos, A.; Salak, A.; Lisenkov, A.; Ferreira, M.; Zheludkevich, M.; Bastos, A.; Lisenkov, A. Influence of preparation conditions of layered double hydroxide conversion films on corrosion protection. *Electrochim. Acta* **2014**, *117*, 164–171. [CrossRef]
53. Evans, D.G.; Slade, R.C. Structural aspects of layered double hydroxides. In *Layered Double Hydroxides*; Duan, X., Evans, D.G., Eds.; Springer: Berlin/Heidelberg, Germany, 2006; pp. 1–87.
54. Kuznetsov, B.; Serdechnova, M.; Tedim, J.; Starykevich, M.; Kallip, S.; Oliveira, M.P.; Hack, T.; Nixon, S.; Ferreira, M.G.S.; Zheludkevich, M. Sealing of tartaric sulfuric (TSA) anodized AA2024 with nanostructured LDH layers. *RSC Adv.* **2016**, *6*, 13942–13952. [CrossRef]
55. Li, Y.; Li, S.; Zhang, Y.; Yu, M.; Liu, J. Enhanced protective Zn–Al layered double hydroxide film fabricated on anodized 2198 aluminum alloy. *J. Alloy. Compd.* **2015**, *630*, 29–36. [CrossRef]
56. Galvão, T.L.; Neves, C.S.; Caetano, A.P.; Maia, F.; Mata, D.; Malheiro, E.; Ferreira, M.J.; Bastos, A.C.; Salak, A.N.; Gomes, J.R.; et al. Control of crystallite and particle size in the synthesis of layered double hydroxides: Macromolecular insights and a complementary modeling tool. *J. Colloid Interface Sci.* **2016**, *468*, 86–94. [CrossRef] [PubMed]
57. Cussler, E.L. *Diffusion: Mass Transfer in Fluid Systems*, 3rd ed.; Cambridge University Press: Cambridge, UK, 2009.
58. Xiang, N.; Song, R.-G.; Zhuang, J.-J.; Song, R.-X.; Lu, X.-Y.; Su, X.-P. Effects of current density on microstructure and properties of plasma electrolytic oxidation ceramic coatings formed on 6063 aluminum alloy. *Trans. Nonferrous Met. Soc. China* **2016**, *26*, 806–813. [CrossRef]

© 2019 by the authors. Licensee MDPI, Basel, Switzerland. This article is an open access article distributed under the terms and conditions of the Creative Commons Attribution (CC BY) license (http://creativecommons.org/licenses/by/4.0/).

Article

Correlation between Defect Density and Corrosion Parameter of Electrochemically Oxidized Aluminum

Hao-Ren Lou [1], Dah-Shyang Tsai [1,*] and Chen-Chia Chou [2]

[1] Department of Chemical Engineering, National Taiwan University of Science and Technology, 43, Keelung Road, Section 4, Taipei 10607, Taiwan; m10606011@mail.ntust.edu.tw
[2] Department of Mechanical Engineering, National Taiwan University of Science and Technology, 43, Keelung Road, Section 4, Taipei 10607, Taiwan; ccchou@mail.ntust.edu.tw
* Correspondence: dstsai@mail.ntust.edu.tw; Tel.: +886-2-2737-6618

Received: 5 December 2019; Accepted: 24 December 2019; Published: 27 December 2019

Abstract: It has been recognized that a connection may exist between defects of oxide coating and its corrosion protection. Such a link has not been substantiated. We prepare two coatings of anodized aluminum oxide (AAO) and plasma electrolytic oxidation (PEO), and analyze them with Mott-Schottky plots and potentiodynamic polarization scans. The as-grown and annealed AAO coatings exhibit both p-type and n-type semiconductor behaviors. Polarization resistance of the AAO coating increases from $(1.8 \pm 1.7) \times 10^8$ to $(4.3 \pm 0.5) \times 10^8$ $\Omega \cdot cm^2$, while corrosion current decreases from $(6.1 \pm 3.6) \times 10^{-7}$ to $(2.3 \pm 0.9) \times 10^{-7}$ $A \cdot cm^{-2}$, as annealing temperature increases from room temperature to 400 °C. The parameter analysis on AAO indicates a positive correlation between corrosion current and donor density, a negative correlation between polarization resistance and donor density. The attempt on correlating corrosion potential gives rise to considerable deviation from a linear fit. The results suggest protection of AAO hinges on its donor density, not acceptor. On the PEO coatings, only the n-type behavior is observed. Intriguingly, the donor density of PEO coating is influenced by the annealing temperature of its pre-anodized layer. The most resistant PEO coating, with pre-anodized and 400 °C annealed AAO, exhibits polarization resistance $(2.1 \pm 0.4) \times 10^9$ $\Omega \cdot cm^2$ and corrosion current $(1.7 \pm 0.4) \times 10^{-8}$ $A \cdot cm^{-2}$.

Keywords: anodized aluminum; corrosion resistance; Mott-Schottky analysis; defect; annealing; plasma electrolytic oxidation

1. Introduction

Electrochemical oxidation offers several value-added attributes to the aluminum surface such as color, hardness, corrosion, and scratch resistances, which enhance its aesthetic and functional purposes. In industrial practice, oxidation of the surface is performed through anodizing or plasma electrolytic oxidation (PEO). Anodizing is commonly carried out in acidic solutions, with an imposed voltage sufficiently low such that electric discharges did not occur. When the imposed voltage is raised to a point that electric discharges emerge and travel on the metal surface, the processing enters the phase of PEO [1–5]. However, the division between anodizing and PEO may not be as clear-cut as described. For example, the treatment of PEO is usually performed in alkaline electrolytic solutions using the constant current mode. The metal surface has to go through a voltage escalating period to reach the state of traveling microdischarges. Thus, PEO is often preceded by a brief period of anodizing.

Anodizing may produce two morphologies of anodic aluminum oxide (AAO): self-ordered porous films and non-porous compact barrier films. In the last two decades, the anodic aluminum oxide composed of regular nanometer pores has attracted tremendous attention, since researchers are in a fervent pursuit of well-defined porous templates that allow them to mold their nanomaterials. The studies on porous-type film have yielded detailed knowledge on how electric current and solution composition can be varied to control the pore size, the interpore spacing, and even the pore diameter in vertical direction [6,7]. On the other hand, a barrier-type film of planar geometry is also desirable. The compact barrier layer with high dielectric strength finds its applications in the electronic devices of metal-insulator-metal capacitor and the microfluidic devices of electrowetting on dielectrics [8,9].

The morphological dissimilarity between porous- and barrier-type films arises from a high-to-low level of incorporating electrolyte anions (highest for regular-pore films, least for non-porous barrier films). In the acidic electrolyte, the incorporating anions could be SO_4^{2-}, $C_2O_4^{2-}$, or PO_4^{3-}. In the neutral or alkaline electrolyte, the anion is the hydroxyl group [6]. Despite the notable morphology differences [10–12], a few aspects are in common. Those oxygen-carrying anions migrate inward, driven by an imposed electric field. Outward migrating Al^{3+} cations diffuse in the opposite direction and both contribute to oxide growth. Both the porous- and the barrier-type films are featured with a barrier layer, which is the oxide adjacent to metallic substrate.

Ion migration in the oxide coating is made possible by point defects. The barrier layer may be viewed as semiconductor because of these frozen point defects [13–16]. When soaked in an electrolytic solution, the barrier layer has two interfaces: the oxide/electrolyte and oxide/metal interfaces. In the AAO literature, researchers concur on the physical picture that one interface is p-type and the other n-type, but cannot agree upon which side is p-type or n-type. The review article of Diggle [17], along with the works of Takahashi [18] and Mibus [19], assumed that the region near oxide/electrolyte interface was anion excess in stoichiometry, therefore, the p-type region. The inner region of oxide/metal interface was excess in metallic cation and n-type behavior. An opposite view was given in the works of Vrublevsky [20–23] and also Benfedda [24], who considered that negative charges such as electrons were trapped at the oxide/electrolyte interface and acted as donors for the n-type behavior. Positive charges, such as holes, were trapped at the oxide/metal interface, acting as acceptors responsible for the p-type behavior.

Researchers have recognized the connection between defects and corrosion and formulated three mechanisms to account for corrosion current via electron and proton conduction [25–27]. Nonetheless, a straightforward correlation of experimental data, to the best of our knowledge, is not reported. Corrosion of the anodized Al alloy 6061 may serve as an excellent example, since 6061 is widely used as structural materials in the aviation and marine industries, and extensively attacked in a chloride containing environment. In this work, we prepare a dense and conformal barrier layer on the 6061 surface. The oxide coating displays the n-type and the p-type behaviors both. We vary the dopant densities with thermal annealing, and show that the corrosion resistance of oxide coating can be correlated to the defect density of donor, not acceptor. Further discussion indicates the external interface of coating is n-type, responsible to the corrosion protection. On the other hand, the inner interface is p-type and is largely related to the growth behavior.

2. Materials and Methods

Two types of coatings on 6061 aluminum alloy were prepared: the AAO coating and the PEO coating with pre-anodized oxide. Anodization of the pre-cleaned surface was performed with a pulsed current of square bipolar waveform in the aqueous solution of 5.5 g·dm^{-3} ammonium pentaborate octahydrate ($NH_4B_5O_8·8H_2O$). The pH value of electrolytic solution was 8.6 and its conductivity 1.45 mS·cm^{-1} at room temperature. The electrical parameters of potentiostatic anodization were set as follows: 200 V in positive polarization and 40 V in negative polarization, 50 Hz in frequency, and 40% in duty ratio. This set of electrical parameters were abbreviated as 200 V (+)/40 V (−), which did not give rise to electrical discharges throughout anodization. The frequency was defined as $(T^+_{on} + T^+_{off} + T^-_{on} + T^-_{off})^{-1}$, in which T^+_{on} and T^-_{on} were the duration periods of positive and negative pulses, respectively. T^+_{off} and T^-_{off} were the resting periods between the positive and negative pulses. The duty ratio was defined as $T^+_{on}/(T^+_{on} + T^+_{off} + T^-_{on} + T^-_{off})$. The anodization period of 200 V (+)/40 V (−) was 20 min. Most of anodized samples were annealed, then subject to the galvanostatic PEO treatment. A few anodized samples were etched in sulfuric acid. The parameters of subsequent

PEO were set 45.3 mA·cm^{-2} (0.7 A) for positive polarization and 51.8 mA·cm^{-2} (0.8 A) for negative polarization, with frequency 50 Hz and duty 40%. This set of PEO parameters was abbreviated as 0.7 A (+)/0.8 A (−). The entire PEO treatment lasted two min and the cell voltage was recorded. In electrochemical oxidations, the sample was mounted at the central position of the electrolytic solution, and the electrical current was sent by a direct current (DC) power supply (DCG-100A, ENI Emerson Electric Co., Saint Louis, MO, USA) with a pulse waveform generator (SPIK-2000A-10H, MELEC, GmbH, Shanghai, China). More details on the setup of anodization and PEO, Figure S1, along with the working procedure, Figure S2, can be found in Supplementary Materials and our previous publication [28].

Annealing of the anodized sample was executed in flowing nitrogen at 100, 150, 250, 300, and 400 °C for one hour using a tubular reactor. Etching of a few anodized samples was performed in 2.0 M sulfuric acid for 15, 30, 45, and 60 min. These samples were subject to Mott-Schottky analysis before etching. The analysis procedure was repeated after etching and washing. For Mott-Schottky analysis, the capacitance in aqueous solutions was recorded every 0.05 V between +2.0 and −2.0 V (vs. Ag/AgCl). The value of space charge capacitance was calculated as the inverse of the imaginary component of impedance at 1 kHz, which was recorded with an electrochemical workstation (Autolab PGSTAT302N, Herisau, Switzerland). Given that double-layer capacitance exceeds space charge capacitance sufficiently and the two are in series, the measured capacitance is dominated by the value of space charge capacitance. The three-electrode setup of impedance measurement involved a solution of 5.5 g·dm^{-3} NH$_4$B$_5$O$_8$ housed in a 500 mL beaker, with the reference electrode of Ag/AgCl (3.0 M KCl) and the counter electrode of stainless mesh. The working electrode of anodized sample was placed at the beaker center, surrounded by the stainless mesh electrode, and was 7.6 cm in diameter. The Ag/AgCl reference was located adjacent to the working electrode.

The surface morphology was examined with a field-emission scanning electron microscope (SEM, JSM-7900F, JEOL, Tokyo, Japan). The surfaces were metallized with platinum prior to SEM observations. The coating thickness was taken as the average value of six different locations of the mounted specimen. Phase analysis was performed with a wide-angle X-ray diffractometer (D2 phaser, Bruker, Billerica, MA, USA), equipped with a CuKα$_1$ radiation source and nickel filter. Diffraction results are plotted in Supplementary Materials. Corrosion resistance was evaluated using the technique of potentiodynamic polarization scan. The measurement was done in a solution of 3.5% sodium chloride at room temperature with a three-electrode setup. The setup involved a working electrode with an exposed area 1.0 cm^2, a reference electrode Ag/AgCl (1.0 M KCl), and a counter electrode of platinum coated titanium mesh 20 mm × 20 mm. Potentiodynamic polarization data were taken using a 1287A electrochemical interface (Solartron Analytical, Leicester, UK). Current data were recorded between −2.0 and 0.0 V at scan rate 5 mV·s^{-1}. The corrosion current (J_{corr}) and the corrosion potential (E_{corr}) were read from the intersection point of anodic and cathodic extrapolated Tafel lines. With the anodic and cathodic Tafel slopes, b_a and b_c, the polarization resistance (R_p) was calculated with the Stern-Geary equation as shown in Equation (1).

$$R_p = \frac{b_a \times b_c}{2.303 \times J_{corr} \times (b_a + b_c)} \qquad (1)$$

3. Results and Discussion

3.1. Defect Density and Corrosion Protection of AAO

Figure 1a,b shows the cross-sectional and top-view images of as-grown barrier layer. As expected, the AAO layer is nonporous and compliant, with minor surface undulations due to of scratches left after polishing. There is no discharge damage found in the oxide since no spark has occurred under 200 V (+)/40 V (−) in the pH 8.6 solution. Thickness of the barrier layer is measured 286 ± 30 nm. The time profiles of positive and negative current, Figure 1c, are consistent with the literature description on self-limiting growth [6]. As time progresses, the positive current density decreases exponentially from an initial value 16.2 mA·cm^{-2} to a steady value 0.65 mA·cm^{-2}. Similarly, the negative current descends from an initial value 4.53 mA·cm^{-2} to a steady one, 0.65 mA·cm^{-2}. The self-limiting growth occurs when the metal piece is made an anode in the electrolytic solution that furnishes oxygen-containing species, and then a film develops uniformly to oppose the ongoing growth since the dielectric film obstructs diffusion of ions, along with their associated charge-transfer reactions. Consequently, holding the voltage constant, the current diminishes with increasing oxide thickness. If the defect concentration of newly-added oxide is the same with that of grown oxide, the barrier layer resistance increases linearly with the layer thickness and the electric current drops exponentially. Hence, the final thickness of barrier layer is largely determined by the imposed voltage, irrelevant to anodization time. The ratio of layer thickness over applied voltage has been reported 1.1–1.4 nm·V^{-1} in literature [9,29], depending on the electrolytic solution. Oxide growth of our 200 V (+)/40 V (−) anodization obeys this rule of thumb, showing a ratio of AAO thickness divided by imposing voltage, ~1.4 nm·V^{-1}.

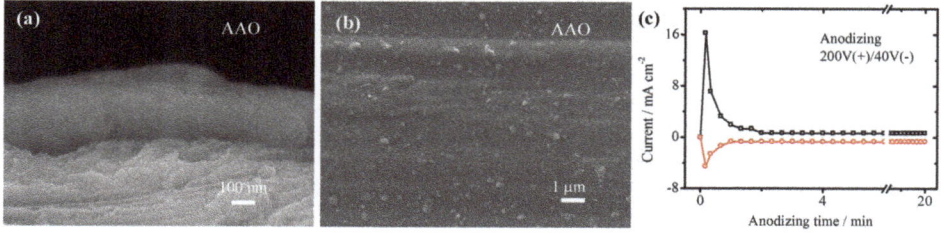

Figure 1. Morphological qualities of the AAO layer and the current density profile in anodizing. The images of (**a**) cross-sectional view and (**b**) top view for the barrier layer grown in the solution of 5.5 g·dm^{-3} NH$_4$B$_5$O$_8$·8H$_2$O. The associated (**c**) anodizing current is performed with the potentiostatic condition 200 V (+)/40 V (−).

Crystal defects provide the essential diffusion channels during growth. After anodization, these remaining defects are vital to the properties of barrier layer. Thermal annealing is known to diminish the defect density of oxide effectively. Figure 2 shows a series of Mott-Schottky plots for the six samples with low-to-high annealing temperature. For each plot of the inverse square of space charge capacitance C_{SC}^{-2} versus electrode potential E, the acceptor density N_a or the donor density N_d can be extracted from the slope of linear segment, as shown in Equations (2) and (3).

$$C_{SC}^{-2} = \frac{-2}{\epsilon\epsilon_0 e N_a}(E - E_{fb} - \frac{kT}{e}), \text{ p-type} \qquad (2)$$

$$C_{SC}^{-2} = \frac{2}{\epsilon\epsilon_0 e N_d}(E - E_{fb} - \frac{kT}{e}), \text{ n-type} \qquad (3)$$

in which ϵ and ϵ_0 denote the relative dielectric constant and the vacuum permittivity, κ is the Boltzmann constant, T is the absolute temperature, e is the electrical charge and E_{fb} is the flat-band potential. For each annealing temperature in Figure 2, one cave-in of the C_{SC}^{-2}-E plot can be found around the electrode potential −0.6 V. Another cave-in may be detected at the more negative potential, related to hydrogen evolution reaction. The cave-in of V-shape indicates the p-type and n-type defects both exist in the barrier layer, since two correlation lines of negative and positive slopes can be drawn. Thus the anodized barrier layer is a p-n heterojunction that may be separated by a neutral region, consistent with the literature. The two slopes generally increase, with increasing annealing temperature. In other words, the acceptor and donor densities decrease. Of the two correlation lines, the intersection potential is assumed to be the flat-band potential value. The E_{fb} value shifts with annealing temperature in the positive direction from −1.4 (as-grown), −0.67 (100 °C), −0.66 (150 °C), −0.53 (250 °C), −0.52 (300 °C), −0.50 V (400 °C), suggesting the defects of barrier layer shift in a systematic manner.

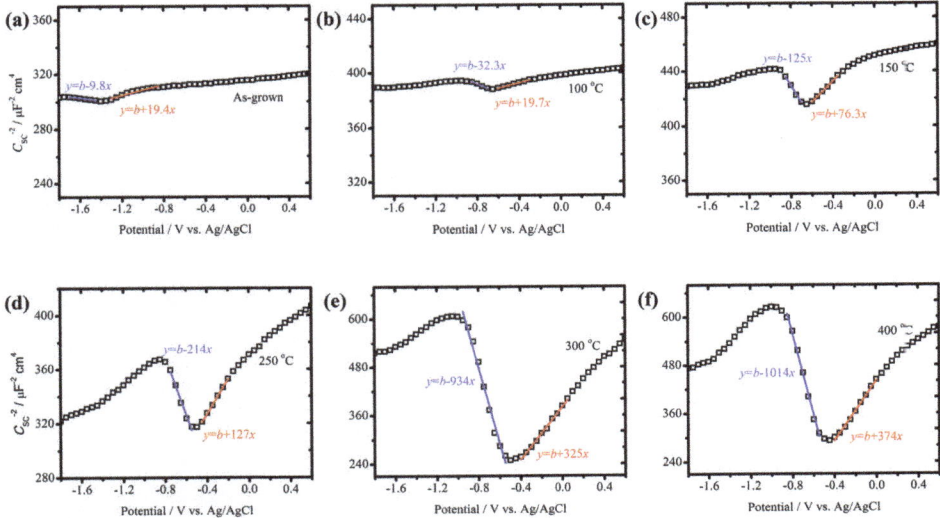

Figure 2. Mott-Schottky plots of annealed AAO layers. The inverse square of space charge capacitance is plotted versus electrode potential between −1.8 and 0.6 V for the (**a**) as-grown barrier layer, the barrier layers after annealing at (**b**) 100, (**c**) 150, (**d**) 250, (**e**) 300, (**f**) 400 °C for 1 h. The positive and negative slopes of linear segment are marked.

Acceptor and donor densities decrease with increasing annealing temperature, as shown in Figure 3. The acceptor density N_a is higher than the donor density N_d in the as-grown barrier layer. Meanwhile, the acceptor density exhibits a higher temperature dependence than the donor density, that is, annealing decreases the acceptor density faster than the donor density. Annealing at 100 °C is sufficient to reduce the acceptor density from 1.7×10^{18} (as-grown) to 5.2×10^{17} cm^{-3}. On the other hand, the donor density decreases slightly from 8.6×10^{17} to 8.4×10^{17} cm^{-3}. The significant drop in donor density occurs at the higher annealing temperature of 150 °C. Further decline in the defect density is less drastic with increasing temperature. The N_a value of 300 °C sample 1.8×10^{16} cm^{-3} is near that of 400 °C, 1.6×10^{16} cm^{-3}. The N_d values of 300 and 400 °C are similar in magnitude as well, with 5.1×10^{16} (300 °C) and 4.4×10^{16} cm^{-3} (400 °C).

Figure 4a presents the typical potentiodynamic polarization curves of annealed barrier layers, in contrast to those of the as-grown barrier layer and the 6061 surface with natural oxide. Comparison of these polarization curves indicates that thermal annealing improves corrosion protection of the barrier layer against 3.5% NaCl solution. Raising the annealing temperature diminishes the corrosion current J_{corr}, raises the polarization resistance R_p, and shifts the corrosion potential E_{corr} in the positive direction. The improvement on anticorrosion appears to be progressive among annealed samples. The most visible enhancement is noted between the as-grown barrier layer and the natural oxide on 6061 surface, yet the overall improvement through annealing is also impressive.

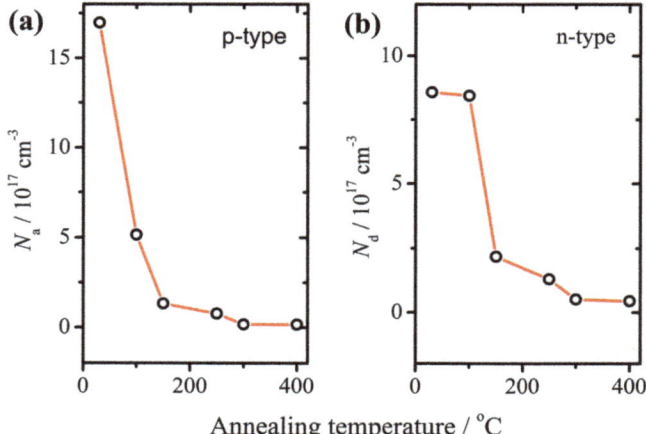

Figure 3. Defect densities of the AAO layer plotted against the annealing temperature. The defect densities of (**a**) acceptor and (**b**) donor decrease with increasing annealing temperature. The annealing temperature of as-grown barrier layer is assigned to be 30 °C.

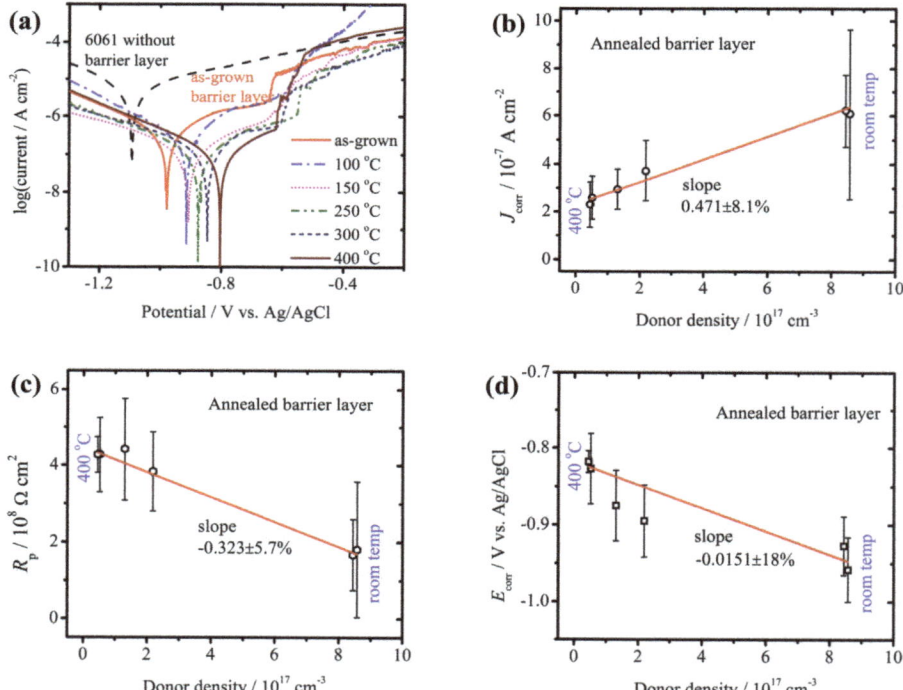

Figure 4. Potentiodynamic polarization curves of the AAO and correlations between corrosion parameters and donor density. (**a**) Potentiodynamic polarization curves at scan rate 5 mV·s^{-1} show the anticorrosion trend. Linear correlation is displayed between (**b**) corrosion current, (**c**) polarization resistance of the AAO layer and its donor density. Considerable deviation is shown in the correlation between (**d**) corrosion potential and donor density.

It is particularly intriguing to attain a linear correlation between corrosion current J_{corr} and donor density N_d, as shown in Figure 4b. Such a linear correlation can also be found between polarization resistance R_p and N_d, Figure 4c. In contrast to J_{corr} and R_p, corrosion potential E_{corr} appears less correlated to N_d, Figure 4d. The above statement is vindicated in the percent deviation of linearity, which is 18% between E_{corr} and N_d, substantially higher than the percent deviation between J_{corr} and N_d, 8.1%, also much higher than the percent deviation between R_p and N_d, 5.7%. The linear correlation is a strong piece of evidence supporting the view that n-type defects are the origin of chloride anion corrosion. We also try to correlate corrosion parameters with acceptor density and obtain a much higher percent deviation of 32% between E_{corr} and N_a, 38% between J_{corr}, and N_a, 38% between R_F and N_a (Figure S3, Supplementary Materials). Evidently, the p-type defects are not the reason of corrosion.

It seems worthwhile to discuss the reason why corrosion potential E_{corr} is less correlated to donor density N_d in contrast to the superior correlation between J_{corr} and N_d, and also that between R_F and N_d. The reported N_d value has been extracted from the linear portion of Mott-Schottky plot, denoting the donor density of the average coating surface. Considering the surface area of our samples, 15.4 cm^2, certain inhomogeneity is bound to exist on the coating. A good correlation between J_{corr} and N_d suggests that the corrosion current is the sum of individual contribution by the typical donor defects, so is R_p and N_d. However, the corrosion potential is not a sum of individual potentials of point defects. The E_{corr} value is more likely influenced by a small group of point defects that do not contribute to corrosion current.

Etching in sulfuric acid offers a different way to identify which side of the barrier layer is n-type. Mott-Schottky plots of the 400 °C-annealed and etched samples are plotted in Figure 5. We note that the C_{SC}^{-2} values of the etched samples are less than those of unetched samples in Figure 2f, indicating that C_{SC} increases in magnitude. In terms of slope decline, the positive slope decreases more than the negative slope does. Hence, the rise in N_d is more pronounced than that in N_a after etching. For example, etching for 15 min, the correlated N_a value rises from 1.6×10^{16} to 8.6×10^{17} cm^{-3}; while the correlated N_d value increases more significantly, from 4.5×10^{16} to 1.8×10^{19} cm^{-3} as shown in Figure 5a. Figure 5b presents the Mott-Schottky plot after 45 min etching, the positive slope is almost zero. The above etching results indicate more donor defects after oxide removal. A plausible explanation is that when the vulnerable oxide being etched, more defects and higher surface heterogeneity are created at the same time.

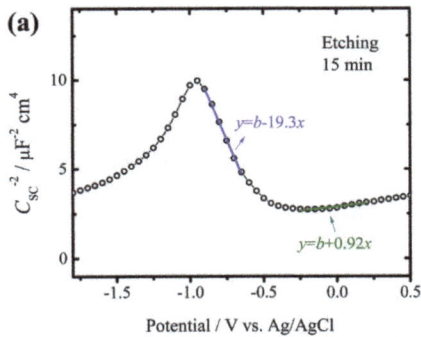
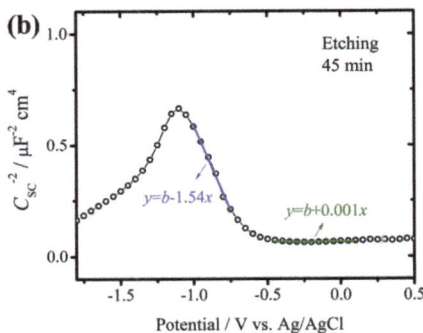

Figure 5. Mott-Schottky plots of the etched AAO layers. The samples of barrier layer have been 400 °C annealed for 1 h are then etched in sulfuric acid for (**a**) 15 and (**b**) 45 min. Note the positive slope flattens as etching time increases.

In view of the preferential etching on n-type oxide, the oxide/electrolyte interface ought to be assigned as the donor interface since the preferential etched surface must directly face the sulfuric acid. Furthermore, this assignment is also supported by the correlation results of Figure 4. Only when the oxide/electrolyte (external) interface is n-type, the corrosion protection could be hinged on the donor density. Otherwise, corrosion of this barrier layer would depend on the acceptor density. Our conclusion on the n-type oxide/electrolyte interface is consistent with the conclusions of Vrublevsky [20–23] and Benfedda [24].

3.2. The PEO Coating with Pre-Grown AAO

The morphological features of the PEO coating are shown in Figure 6. The sample has been pre-anodized under 200 V (+)/40 V (−), then micro-arc treated at 0.7 A (+)/0.8 A (−). The coating thickness is 760 ± 35 nm, Figure 6a, much thicker than its pre-anodized barrier layer, considering the PEO period is 2 min only. Electrical discharges of PEO leave scars, which may be elusive in the cross-sectional image of the coating, yet distinct in the top-view image as shown in Figure 6b. The coating surface shows many pin holes of submicron size associated with a terrain of frozen lava swellings, much different from the pre-anodized surface. Also encircled in red are a few micron-sized grains of hexagonal facet, which might be the precursor of alpha alumina. The crystalline phase of the PEO coating is gamma aluminum oxide, as indicated in its X-ray diffraction result (Figure S4, Supplementary Materials).

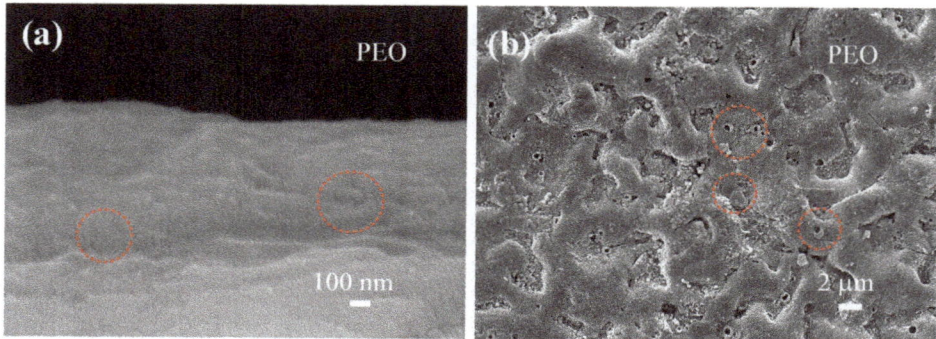

Figure 6. Morphology of the PEO coating with pre-anodized oxide. The SEM images of (**a**) cross-sectional view and (**b**) top view of the PEO coating, which is preceded by an AAO layer prepared at 200 V (+)/40 V (−). Large circles mark the electrical discharge damages.

Mott-Schottky plots of two PEO samples are shown in Figure 7, indicating only the n-type behavior is detected. One PEO sample has been treated with a pre-anodized layer without annealing, Figure 7a, the other with a pre-anodized layer being 250 °C-annealed, Figure 7b. Annealing temperature can affect the donor density of the PEO coating. Without annealing, the donor density is 2.92×10^{17} cm^{-3}. The donor density of PEO coating decreases with increasing pre-annealing temperature, with 2.67×10^{17} (150 °C), 2.37×10^{17} (250 °C), and 1.95×10^{17} cm^{-3} (400 °C).

Since the extent of N_d decline in PEO coating is narrow, its influences on corrosion protection are moderate. Figure 8 shows how much corrosion protection is affected by the donor density of PEO coating. Figure 8a indicates the corrosion potential E_{corr} increases with increasing annealing temperature of the AAO layer, −0.77 ± 0.055 V (without annealing), −0.74 ± 0.03 V (250 °C annealed AAO), −0.72 ± 0.02 V (400 °C annealed AAO). The corrosion current J_{corr} of PEO coating decreases with increasing annealing temperature; $(3.3 ± 0.9) \times 10^{-8}$ (AAO without annealing), $(1.7 ± 0.5) \times 10^{-8}$ (250 °C annealed AAO), $(1.7 ± 0.4) \times 10^{-8}$ A·cm^{-2} (400 °C annealed AAO). And the polarization resistance R_p increases with increasing annealing temperature; $(1.1 ± 0.3) \times 10^{9}$ (AAO without annealing), $(2.3 ± 0.8) \times 10^{9}$ (250 °C annealed AAO), $(2.1 ± 0.4) \times 10^{9}$ Ω cm^2 (400 °C annealed AAO).

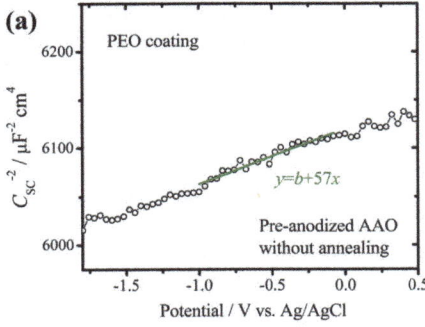
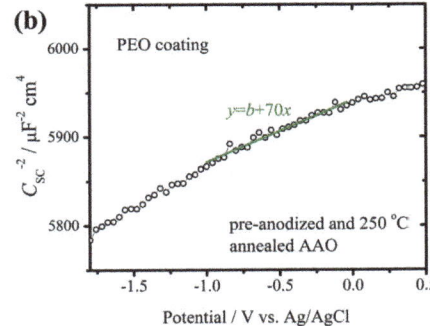

Figure 7. Typical Mott-Schottky plots of the PEO coatings. The plots of C_{SC}^{-2} versus electrode potential are drawn for the PEO coatings with an AAO layer (**a**) without annealing, (**b**) 250 °C annealing.

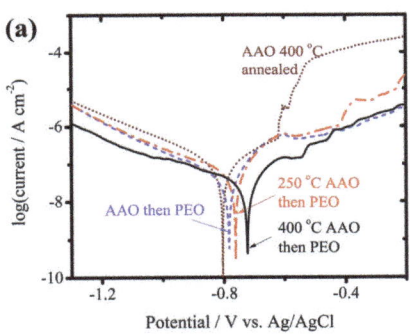
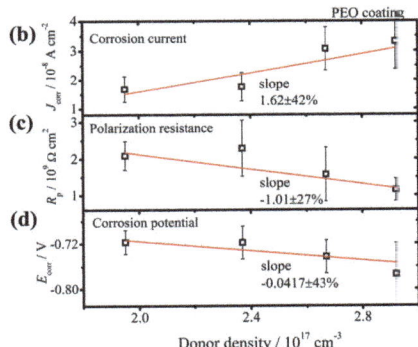

Figure 8. Influences of the AAO layer on anticorrosion of following PEO coating. Comparison of potentiodynamic polarization curves at scan rate 5 mV·s^{-1} shows (**a**) the influences of annealing temperature on anticorrosion of its PEO coating. Correlation between donor density and (**b**) corrosion current, (**c**) polarization resistance, (**d**) corrosion potential of the PEO coating.

We ought to mention that linear correlation of the PEO coating contains substantial uncertainty inferior to that of the AAO layer. Correlation between corrosion protection and donor density of PEO coatings, the percent deviation is 27% between R_p and N_d, 42% between J_{corr} and N_d, 43% between E_{corr} and N_d. We attribute the weak correlation to the damaging effects of electrical discharge. Extra thickness of the PEO coating provides additional protection, but its porosity also brings more uncertainty.

It is of interest to look into why the donor density of PEO coating is affected by annealing temperature of its pre-anodized AAO. One plausible cause is defects of the pre-anodized AAO layer alter the voltage profile of PEO treatment. Figure 9a contrasts the voltage-time profiles of 400 and 250 °C annealed samples, with that of the sample without annealing. The voltage profiles of 400 and 250 °C annealing are consistently higher than that of the one without annealing. Thus, a higher positive voltage is required for the AAO layer of less defects to reach the preset current density. If the pre-anodized layer is partially etched away, the remaining oxide affects the voltage-time profile of PEO differently. Figure 9b compares the voltage profile of the pre-anodized sample without etching with those of the etched samples. As a whole, the positive voltage of the samples etched for 15, 30, 45, and 60 min begins with a lower value than the unetched sample. However, the voltage of etched samples surpasses that of the unetched sample after 5 or 9 s. We generally observe the first spark at 10 s (Video S1, Supplementary Materials), therefore, the E_{10S} value is physically meaningful. The E_{10S} values in Figure 9b are listed as: 507 (15 min), 475 (30 min), 461 (45 min), 437 (60 min), and 421 V (unetched). An etched sample seems to develop a coating with higher dielectric strength than the unetched sample after several PEO seconds, so that its positive voltage is higher. Therefore, defects of the pre-anodized oxide exert influences on its following PEO coating. Compared with the unetched sample, the etched AAO layer grows faster, so does the positive voltage when PEO time is between 2 and 18 s. The acceptor defects and diminished outer layer seem to be responsible for the fast growth in this brief period.

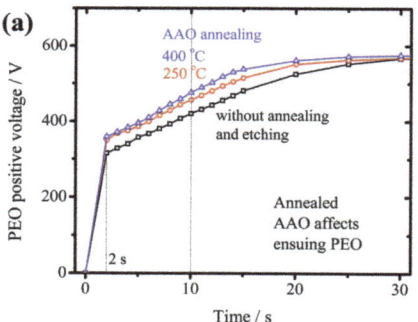
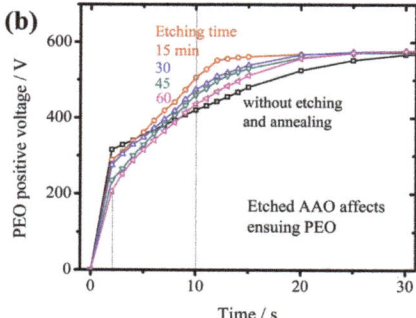

Figure 9. Positive voltage evolution in the initial stage of PEO treatment. Comparison of the voltage-time profiles of PEO that began with (**a**) 400 and 250 °C annealed barrier layer, and the barrier layer without annealing. Comparison of the voltage-time profiles of PEO that started with (**b**) the barrier layers of 15, 30, 45, and 60 min etching and the barrier layer without etching and annealing.

4. Conclusions

Point defects are known to affect corrosion protection properties in various manners, but researchers seldom connect the two directly. In this research, the point defect densities of acceptor and donor are varied through annealing and etching of the AAO layer, and the anticorrosion parameters of the AAO layer are measured with potentiodynamic polarization scans. We find that a linear correlation between corrosion current and donor density, also between polarization resistance and donor density. The correlation between corrosion potential and donor density can be fitted with considerable deviation. For PEO coatings, only the n-type behavior is observed. Similar correlations can be established between the corrosion properties and donor density for PEO coating. However, deviations from linear correlation are substantial, since the electrical discharges during PEO introduce porosity and uncertainty in anticorrosion properties.

Supplementary Materials: The following are available online at http://www.mdpi.com/2079-6412/10/1/20/s1, Figure S1: Schematic diagram of setup for anodizing the 6061 sample, following by PEO treatment, Figure S2: Experimental procedure of electrochemical oxidation, characterization, and corrosion analysis, Figure S3: Linear correlation between corrosion parameters and acceptor density. In contrast to Figure 4b–d, the correlations are poor with respect to acceptor density. Correlation is attempted between (a) J_{corr} and N_a, (b) R_p and N_a, (c) E_{corr} and N_a, Figure S4: XRD patterns of AAO and PEO coatings. (a) The result of 400 °C annealed barrier layer displays the diffraction lines of metallic aluminum substrate only. (b) The diffraction result of PEO coating that began with an AAO precursor shows the gamma aluminum oxide features in addition to the diffraction lines of aluminum metal, Video S1: The first appearance of microdischarges in PEO.

Author Contributions: Conceptualization, formal analysis and writing, D.-S.T.; investigation and validation, H.-R.L.; resources, C.-C.C. All authors have read and agreed to the published version of the manuscript.

Funding: The authors would like to thank Ministry of Science and Technology of Taiwan for financial support of this work through the project MOST-106-2221-E-011-119-MY3.

Acknowledgments: We would like to thank Ching-Hwa Ho of NTUST for allowing the oscilloscope and other electronic measurement instruments in this work.

Conflicts of Interest: The authors declare no conflict of interest.

References

1. Clyne, T.W.; Troughton, S.C. A review of recent work on discharge characteristics during plasma electrolytic oxidation of various metals. *Int. Mater. Rev.* **2018**, *63*, 127–164. [CrossRef]
2. Liu, C.Y.; Tsai, D.S.; Wang, J.M.; Tsai, J.T.J.; Chou, C.C. Particle size influences on the coating microstructure through green chromia inclusion in plasma electrolytic oxidation. *ACS Appl. Mater. Interfaces* **2017**, *9*, 21864–21871. [CrossRef] [PubMed]
3. Tsai, D.S.; Chen, G.W.; Chou, C.C. Probe the micro arc softening phenomenon with pulse transient analysis in plasma electrolytic oxidation. *Surf. Coat. Technol.* **2019**, *357*, 235–243. [CrossRef]
4. Matykina, E.; Arrabal, R.; Skeldon, P.; Thompson, G.E.; Belenguer, P. AC PEO of aluminum with porous alumina precursor films. *Surf. Coat. Technol.* **2010**, *205*, 1668–1678. [CrossRef]
5. Matykina, E.; Arrabal, R.; Mohamedm, A.; Skeldon, P.; Thompson, G.E. Plasma electrolytic oxidation of pre-anodized aluminum. *Corros. Sci.* **2009**, *51*, 2897–2905. [CrossRef]
6. Lee, W.; Park, S.J. Porous anodic aluminum oxide: Anodization and templated synthesis of functional nanostructures. *Chem. Rev.* **2014**, *114*, 7487–7556. [CrossRef]
7. Thompson, G.E. Porous anodic alumina: Fabrication, characterization and applications. *Thin Solid Films* **1997**, *297*, 192–201. [CrossRef]
8. Hourdakis, E.; Koutsoureli, M.; Papaioannou, G.; Nassiopoulou, A.G. Leakage current and charging discharging processes in barrier-type anodic alumina thin films for use in metal-insulator-metal capacitors. *J. Appl. Phys.* **2018**, *123*, 215301. [CrossRef]
9. Mibus, M.; Jensen, C.; Hu, X.; Knospe, C.; Reed, M.L.; Zangari, G. Dielectric breakdown and failure of anodic aluminum oxide films for electrowetting systems. *J. Appl. Phys.* **2013**, *114*, 014901. [CrossRef]
10. Capraz, O.O.; Overmeere, Q.V.; Shrotriya, P.; Herbert, K.R. Stress induced by electrolyte anion incorporation in porous anodic aluminum oxide. *Electrochim. Acta* **2017**, *238*, 368–374. [CrossRef]
11. Capraz, O.O.; Shrotriya, P.; Skelton, P.; Thompson, G.E.; Herbert, K.R. Factors controlling stress generation during the initial growth of porous anodic aluminum oxide. *Electrochim. Acta* **2015**, *159*, 16–22. [CrossRef]
12. Dou, Q.; Overmeere, Q.V.; Shrotriya, P.; Li, W.; Herbert, K.R. Stress induced by incorporation of sulfate ions into aluminum oxide films. *Electrochem. Commun.* **2018**, *88*, 39–42. [CrossRef]
13. Stojadinovic, S.; Vasilic, R.; Nedic, Z.; Kasalica, B.; Belca, I.; Zekovic, L. Photoluminescent properties of barrier aluminum oxide films on aluminum. *Thin Solid Films* **2011**, *519*, 3516–3521. [CrossRef]
14. Nigo, S.; Kubota, M.; Harada, Y.; Hirayama, T.; Kato, S.; Kitazawa, H.; Kido, G. Conduction band caused by oxygen vacancies in aluminum oxide for resistance random access memory. *J. Appl. Phys.* **2012**, *112*, 033711. [CrossRef]
15. Chang, J.K.; Liao, C.M.; Chen, C.H.; Tsai, W.T. Microstructure and electrochemical characteristics of aluminum anodized film formed in ammonium adipate solution. *J. Electrochem. Soc.* **2003**, *150*, B266–B273. [CrossRef]
16. Sousa, C.T.; Leitao, D.C.; Proenca, M.P.; Apolinario, A.; Correia, J.G.; Ventura, J.; Araujo, J.P. Tunning pore filling of anodic alumina templates by accurate control of the bottom barrier layer thickness. *Nanotechnology* **2011**, *22*, 315602. [CrossRef]
17. Diggle, J.W.; Downie, T.C.; Goulding, C.W. Anodic oxide films on aluminum. *Chem. Rev.* **1969**, *69*, 365–405. [CrossRef]
18. Takahashi, H.; Kasahara, K.; Fujiwara, K.; Seo, M. The cathodic polarization of aluminum covered with anodic oxide films in a neutral borate solution. *Corros. Sci.* **1994**, *36*, 677–688. [CrossRef]
19. Mibus, M.; Jensen, C.; Hu, X.; Knospe, C.; Reed, M.L.; Zangari, G. Improving dielectric performance in anodic aluminum oxide via detection and passivation of defect states. *Appl. Phys. Lett.* **2014**, *104*, 244103. [CrossRef]
20. Vrublevsky, I.; Jagminas, A.; Schreckenbach, J.; Goedel, W.A. Electronic properties of electrolyte/anodic alumina junction during porous anodizing. *Appl. Surf. Sci.* **2007**, *253*, 4680–4687. [CrossRef]
21. Vrublevsky, I.; Jagminas, A.; Schreckenbach, J.; Goedel, W.A. Embedded space charge in porous alumina films formed in phosphoric acid. *Electrochim. Acta* **2007**, *53*, 300–304. [CrossRef]
22. Vrublevsky, I.; Parkoun, V.; Schreckenbach, J.; Goedel, W.A. Dissolution behavior of the barrier layer of porous oxide films on aluminum formed in phosphoric acid studied by a re-anodizing technique. *Appl. Surf. Sci.* **2006**, *252*, 5100–5108. [CrossRef]

23. Vrublevsky, I.; Parkoun, V.; Sokol, V.; Schreckenbach, J. Study of chemical dissolution of the barrier oxide layer of porous alumina films formed in oxalic acid using a re-anodizing technique. *Appl. Surf. Sci.* **2004**, *236*, 270–277. [CrossRef]
24. Benfedda, B.; Hamadou, L.; Benbrahim, N.; Kadri, A.; Chainet, E.; Charlot, F. Electrochemical impedance investigation of anodic alumina barrier layer. *J. Electrochem. Soc.* **2012**, *159*, C372–C381. [CrossRef]
25. Hassel, A.W.; Lohrengel, M.M. Initial stages of cathodic breakdown of thin anodic aluminum oxide films. *Electrochim. Acta* **1995**, *40*, 433–437. [CrossRef]
26. Hassel, A.W.; Diesing, D. Modification of trap distributions in anodic aluminum tunnel barriers. *J. Electrochem. Soc.* **2007**, *154*, C558–C561. [CrossRef]
27. Schultze, J.W.; Lohrengel, M.M. Stability, reactivity and breakdown of passive films. *Electrochim. Acta* **2000**, *45*, 2499–2513. [CrossRef]
28. Yeh, S.C.; Tsai, D.S.; Wang, J.M.; Chou, C.C. Coloration of aluminum alloy surface with dye emulsions while growing a plasma electrolytic oxide layer. *Surf. Coat. Technol.* **2016**, *287*, 61–66. [CrossRef]
29. Hunter, M.S.; Fowle, J. Determination of barrier layer thickness of anodic oxide coatings. *J. Electrochem. Soc.* **1954**, *101*, 481–485. [CrossRef]

© 2019 by the authors. Licensee MDPI, Basel, Switzerland. This article is an open access article distributed under the terms and conditions of the Creative Commons Attribution (CC BY) license (http://creativecommons.org/licenses/by/4.0/).

Article

Production of Phosphorescent Coatings on 6082 Aluminum Using $Sr_{0.95}Eu_{0.02}Dy_{0.03}Al_2O_{4-\delta}$ Powder and Plasma Electrolytic Oxidation

Krisjanis Auzins [1], Aleksejs Zolotarjovs [1], Ivita Bite [1], Katrina Laganovska [1], Virginija Vitola [1], Krisjanis Smits [1,*] and Donats Millers [1,2]

[1] Institute of Solid State Physics, University of Latvia, Kengaraga str. 8, LV-1063 Riga, Latvia; krisjanis.auzins@gmail.com (K.A.); aleksejs.zol@gmail.com (A.Z.); ivita.bite@gmail.com (I.B.); katrina.laganovska@gmail.com (K.L.); virgiinija@gmail.com (V.V.); dmillers@latnet.lv (D.M.)
[2] ElGoo Tech Ltd., Peldu str. 7, LV-3002 Jelgava, Latvia
* Correspondence: smits@cfi.lu.lv; Tel.: +371-26-538-386

Received: 28 October 2019; Accepted: 11 December 2019; Published: 16 December 2019

Abstract: In this study, a new approach for producing phosphorescent aluminum coatings was studied. Using the plasma electrolytic oxidation (PEO) process, a porous oxide coating was produced on the Al6082 aluminum alloy substrate. Afterwards, activated strontium aluminate ($SrAl_2O_4$: Eu^{2+}, Dy^{3+}) powder was filled into the cavities and pores of the PEO coating, which resulted in a surface that exhibits long-lasting luminescence. The structural and optical properties were studied using XRD, SEM, and photoluminescence measurements. It was found that the treatment time affects the morphology of the coating, which influences the amount of strontium aluminate powder that can be incorporated into the coating and the resulting afterglow intensity.

Keywords: plasma electrolytic oxidation (PEO); aluminum 6082; luminescent coatings; phosphorescence

1. Introduction

Persistent luminophores or "phosphors" are materials that exhibit long-lasting luminescence that is often called phosphorescence, and these have been known and widely used for more than a century in various glow-in-the-dark objects. During the 20th century, the most commonly used persistent luminophore was zinc sulfide (ZnS) in combination with a suitable activator ion (e.g., copper, cobalt) [1]. However due to the comparatively low intensity and duration of persistent luminescence, the use of this persistent luminophore has decreased in favor of strontium aluminate-based persistent luminophores. Discovered in the mid-20th century, strontium aluminate [2] activated with divalent europium ($SrAl_2O_4$: Eu^{2+}) has since become one of the most widely used persistent luminophores. With the addition of Dy^{3+} in the matrix in the 1990s [3], the afterglow time increased significantly without the use of any radioactive materials. Compared to ZnS-persistent luminophores, it has superior luminescence intensity, and its afterglow duration can reach up to 20 h [4]. There are various methods for synthesizing strontium aluminates (commonly in the form of powder), the most popular being the solid-state method [5], combustion synthesis [6], the sol–gel method [7], and the precipitation method [8].

Long-lasting luminescence materials mostly find applications in places where lighting is necessary in case of a power failure such as emergency exit aisles in buildings or public transportation vehicles. Safety signs, road markings [9], and parts of glow-in-the-dark objects such as watch arms, remote control buttons, toys, and other everyday items are another domain in which long persistent luminescence materials are used. At the moment, most of the persistent luminescence objects are created by applying phosphorescent paint to the desired objects, which a lot of the time includes metallic surfaces, e.g., safety signs. With time, the phosphorescent paint can start to wear off and lose its brightness; thus, a more durable method for applying luminescent coatings to different surfaces should be developed.

Plasma electrolytic oxidation (PEO), also known as micro-arc oxidation (MAO), is one of the newest and most studied methods for treating valve metal surfaces, which creates a porous ceramic oxide layer on the metal surface and is most commonly used on metals such as Mg [10], Al [11], Ti [12], and Zr [13]. These metals naturally develop only a thin oxide layer of a few nanometers, which does not protect the surface of the metal from mechanical damage; so, it is necessary to create an additional protective layer. Although PEO phenomena have been known for around 100 years, more rigorous research on this method was begun in the late 1990s by Yerokhin [14]. While similar to conventional anodization, the main difference of PEO is the use of high voltages, which are usually above 400 V. During the beginning phases of treatment, a dielectric oxide layer is grown, and at some point, dielectric breakdown occurs. This leads to local high-energy spark discharges through the dielectric layer. These discharges cause small channels to form on the coating, which results in the increased porosity of the coating [15]. The morphology and properties of the PEO coating can be modified by changing the voltage and current density as well as the polarity and pulse duration [16].

Metal alloys such as aluminum or magnesium are widely used in automotive [17], aerospace [18], and aviation industries because of their light weight and ease of workability. Despite being mostly used for enhancing the wear and corrosion resistance properties of metals [19,20], the PEO method has proven to be useful for creating functionalized coatings, and in the last few years, there has been a growing amount of research devoted to studying these prospects. PEO has been used for producing photocatalytic [21], gas sensing [22], and dosimetric [23] coatings as well as coatings for biomedical applications such as dental [24] and orthopedic implants [25]. One of the more recent fields of study has been PEO application for the development of luminescent coatings. Several articles report the successful incorporation of rare earth luminescent ions in the coatings of different metals [26,27].

One of the latest innovations is the development of long-lasting luminescence coatings using the PEO process. The combination of the protective properties of a PEO coating and long-lasting luminescence opens up possibilities for different practical applications. By adding raw materials in powder form to the electrolyte, it was possible to synthesize strontium aluminate luminophore on an aluminum substrate in a single-step PEO process [28]. However, the method has a few disadvantages that need to be addressed. Large amounts of lanthanide oxides need to be mixed in the electrolyte in order to produce a coating with long afterglow, which significantly increases costs. Another problem is the aging or degradation of the electrolytes, which leads to a decreased quality of the coating, as it affects the characteristics of discharges and the porosity of the PEO coating [29]. Lastly, the quantum yield of such coatings compared to commercially available activated strontium aluminate powder is still relatively low. Although the method of developing long-lasting luminescence coatings is very promising for various practical applications, the shortcomings of it must be overcome, and different approaches should be studied.

In this article, a new method of filling the cavities of PEO coatings with commercially available activated strontium aluminate powder to produce phosphorescent aluminum oxide coatings is developed and studied. Such a method would decrease the amount of raw material consumption and possibly increase the luminescence quantum yield of long-lasting luminescence coatings.

2. Materials and Methods

2.1. Materials

Commercial aluminum alloy Al6082 was used as a substrate material (Al: 95.2%–98.3%; Cr: 0.25% max; Cu: 0.1% max; Fe: 0.5% max; Mg: 0.6 to 1.2%; Mn: 0.4% to 1.0%; Si: 0.7%–1.3%; Ti: 0.1% max; Zn: 0.2% max; residuals: 0.15% max). The aluminum was cut using a mechanical cutting blade to form specimens in 60 mm × 15 mm × 3 mm sizes with a total area of 22.5 cm^2. The electrolytes consisted of 600 mL of deionized water (conductivity 0.0555 µSm/cm) and 1.0 g/l KOH ("RK Chem"). Commercial Sigma Aldrich strontium aluminate powder doped with europium, and dysprosium ions ($Sr_{0.95}Eu_{0.02}Dy_{0.03}Al_2O_4$, purity ≥ 99%) were used for pore-filling purposes.

2.2. Synthesis

For the production of the PEO coatings, a custom-built power supply unit BS4000 (A5V1000/300) by Elgoo Tech Ltd (Jelgava, Latvia) was used. The power output of the unit is up to 5 kW with a voltage limit at 1000 V and a current limit at 5 A. The unit is controlled via a computer interface to set up the desired voltage and current parameters. All the samples were prepared using a non-pulsed direct current with 700 V and 3 A limiting parameters. The surface area covered during PEO treatment was 16.65 cm^2, which yielded a current density of 0.18 A/cm^2. Before the PEO process, aluminum substrates were washed with deionized water and acetone. The container for the PEO process is made of double-walled glass with inlets for water cooling. To ensure sufficient cooling, the electrolyte was constantly stirred using a magnetic stirrer. During the PEO process, large amounts of heat are dissipated, and a part of the electrolyte evaporates. To keep the amount of electrolyte unchanged, constant refilling is necessary. After the PEO process, samples were washed with deionized water and dried in air at room temperature. Afterwards, approximately 0.1 g of commercial activated strontium aluminate powder was mixed with ethanol and milled in an agate mortar with a pestle for 5 min. Then, the mixture was applied to PEO coatings using a pipette and using the tip of the pipette equally distributed along the surface of the sample. Then, the sample was put in a furnace at 80 °C for 5 min to evaporate the ethanol. When the powder had dried, the sample was laid on a firm surface, and another similarly sized non-treated aluminum sheet was carefully placed on top of it, and constant pressure (approx. 1 kg/cm^2) was applied to the system for 5 min to ensure that the powder was pressed into the pores. Afterwards, the sample was washed thoroughly with deionized water to flush away any particles that did not adhere well enough to the surface. In total, 11 PEO samples were synthesized using different treatment durations and studied during this research. The parameters for all samples are shown in Table 1. A graphical step-by-step process representation is shown in Figure 1.

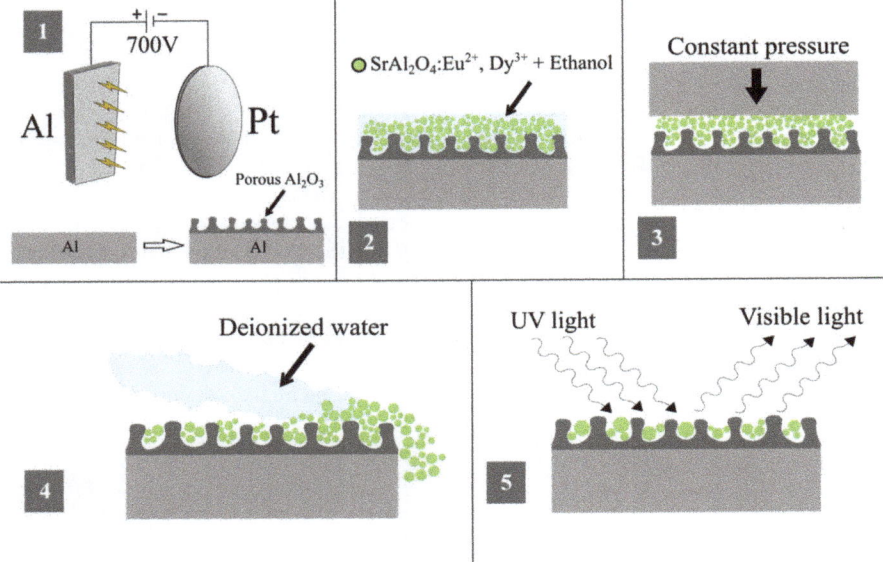

Figure 1. Graphical step-by-step process representation.

Table 1. Plasma electrolytic oxidation (PEO) sample synthesis parameters.

Sample No.	Voltage, V	Current Density, A/cm^2	Duration, Min	Powder Filling
P15N			15	
P20N			20	
P30N	700	0.18	30	No
P40N			40	
P45N			45	
P60N			60	
P15Y			15	
P20Y			20	
P30Y	700	0.18	30	Yes
P40Y			40	
P45Y			45	
P60Y			60	

2.3. Analysis Methods

Sample morphology and phase composition were studied using scanning electron microscopy (SEM) and X-ray diffraction (XRD). SEM images were taken using a Phenom-World Phenom Pro scanning electron microscope registering backscattered electron (BSE) images with 10 kV acceleration voltage. Cross-section images were taken using an SEM Tescan Lyra equipped with an energy dispersive X-ray spectrometer (EDX) operated at 15 kV. X-ray diffraction patterns were acquired using a Rigaku MiniFlex 600 X-ray diffractometer using a cathode voltage of 40 kV and current of 15 mA with Cu Kα radiation (1.5418 Å). The step size for the XRD measurements was 0.05 2θ with a scan speed of 10 2θ/min. Photoluminescence (PL) spectra measurements were made using an Andor Shamrock B 303i spectrograph in combination with an Andor DU-401A-BV CCD camera. The excitation source for PL measurements was a CryLas Nd:YAG laser (266 nm) (spot size on sample: approximately 3 mm in diameter). Luminescence decay kinetics were measured using a Horiba iHR320 monochromator coupled with a Hamamatsu R928P photomultiplier tube using a 5 mW 405 nm diode laser as the excitation source.

3. Results and Discussion

The pore-filling method developed and used during this study allows the production of long afterglow phosphorescent PEO coatings. The naming of this method may be debated, since the powder is filled in the cavities or cracks of the PEO coating rather than pores, but the idea of filling up empty space in the coating remains. The produced samples can be seen in Figure 2. Visually, the appearance of the coating does not differ before and after filling the pores. During normal conditions, the coating is light gray and does not exhibit any luminescence properties. When the sample is illuminated by UV light, the electrons from Eu^{2+} are transferred to the traps, and after ceasing the excitation, the electrons from traps recombine with Eu^{3+} [30,31], and the sample exhibits a bright green long afterglow visible with the naked eye, which is characteristic to the $SrAl_2O_4$: Eu^{2+}, Dy^{3+} luminophore. The phosphorescent coating has good adhesion to the aluminum alloy surface as it was obtained during electrochemical oxidation, and it is also to some extent water-resistant, although strontium aluminate is partly soluble in water, and the luminescent coating might degrade over time without additional protection.

3.1. Structure and Morphology

The XRD patterns of the acquired PEO coatings before and after pore filling are shown in Figure 3. The samples mainly contain γ-Al_2O_3 phase alumina and small traces of α-Al_2O_3 phase. Al peaks from the substrate are also visible in the pattern. The sample after pore filling additionally contains peaks that correspond to $SrAl_2O_4$. The detected phases in all the samples are presented in Table 2. While the γ-Al_2O_3 phase can be detected in all the samples, the α-Al_2O_3 is only visible in samples with a treatment duration longer than 30 min, although some samples that exceed the 30-min duration still do not show an α-Al$_2$O$_3$ phase. This is due to the presence of an α-Al$_2$O$_3$ phase in a small quantity, and therefore, it is right on the detection limit of the XRD device. The SrAl$_2$O$_4$ phase is detected only in samples with longer PEO durations, i.e., 40 and 45-min treatments. This is because the longer treatment time allows more powder to be filled into the coating, which can be explained by analyzing SEM images. Although samples P15Y, P20Y, and P30Y have the luminescent powder in them (as confirmed by luminescence measurements later), the content is not enough to be detected by XRD.

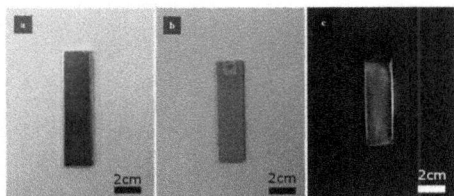

Figure 2. Sample (**a**) before PEO (P45N), (**b**) after PEO and pore filling (P45Y), and (**c**) after excitation with UV light.

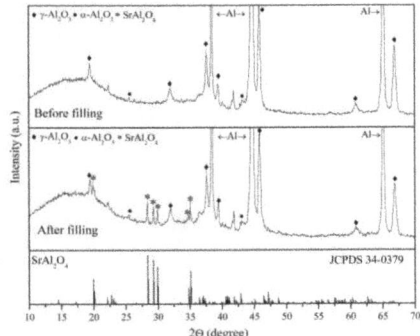

Figure 3. XRD patterns of 45-min sample before and after pore filling.

Table 2. Detected phases by XRD analysis in samples.

Sample Name	γ-Al$_2$O$_3$	α-Al$_2$O$_3$	SrAl$_2$O$_4$
P15N	X	-	-
P20N	X	-	-
P30N	X	-	-
P40N	X	X	-
P45N	X	X	-
P60N	X	X	-
P15Y	X	-	-
P20Y	X	-	-
P30Y	X	X	-
P40Y	X	X	X
P45Y	X	X	X
P60Y	X	X	-

The SEM images displaying morphology of the coatings after different PEO treatment times are shown in Figure 4. It can be seen that longer PEO treatment duration develops thicker and more densely packed cavities and pores. After 15 and 20 min (Figure 4a,b), the pores are very small with a size of around 1 micrometer. After 30 min or a longer time of treatment (Figure 4c–e), the pores and cavities are larger (2–10 micrometers) and denser.

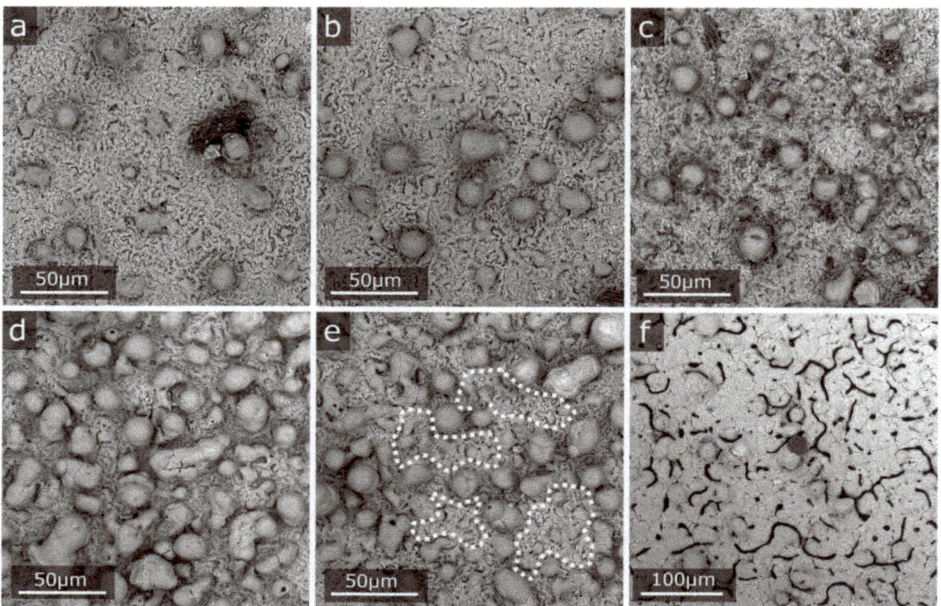

Figure 4. SEM images of samples before pore filling (**a**) 15 min – P15N; (**b**) 20 min – P20N; (**c**) 30 min – P30N; (**d**) 40 min – P40N; (**e**) 45 min – P45N (cavities are marked with dashed line), and (**f**) 60 min – P60N.

The optimal PEO time was found to be 45 min, as continuing the treatment longer leads to the transition of the coating to a different, less porous structure. The effect is shown in Figure 4f for P60N. Beginning from the sides of the sample, the plasma discharges and gradually becomes less dense, and the morphology transforms to one without large pores and containing thin long cavities. After approximately 60 min, most of the sample morphology had transformed (P60N). This was found to reduce the amount of powder that can be filled into the coating significantly. Figure 5a displays commercial strontium aluminate powder. It can be seen that the average grain size is around 30–50 micrometers. To facilitate the incorporation of powder into the coating, milling was necessary. Figure 5b shows strontium aluminate powder after milling for 5 min. The majority of grains after milling are smaller than 10 micrometers. Sieving was not performed after the milling. SEM images with powder filled into the coating are shown in Figure 5c. The particles build into the cavities that are formed from round alumina structures. In Figure 5d, the surface is shown at a smaller magnification, which represents how densely the powder has filled up the pores of the sample with 45-min PEO treatment.

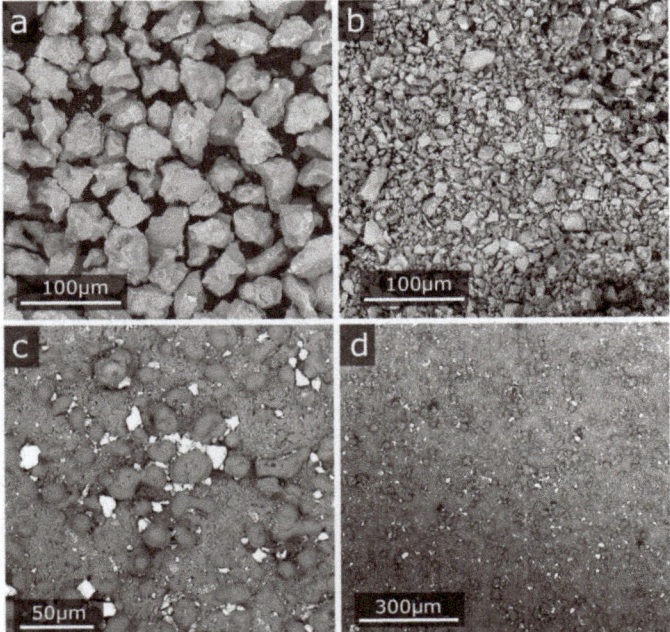

Figure 5. (**a**,**b**) Strontium aluminate powder before and after milling; (**c**,**d**) P45Y sample after pore filling

To analyze the incorporation of powder in the pores and evaluate the thickness of the coating, a cross-section image of the P45Y sample was taken (Figure 6). As one can see, the thickness of the coating varies greatly due to the high porosity of the alumina, and the average thickness is measured at approximately 40 µm. Moreover, evaluation of the composition of the coating gives a confirmation of the presence of strontium aluminate powder in the pores. An interesting observation can be made: the Mg atoms are also detected in both the aluminum alloy (Mg is the main alloying element in Al6082 alloy) and in the coating (taken from the alloy itself and incorporated in the structure of the alumina). Other elements are below the detection limit of the setup.

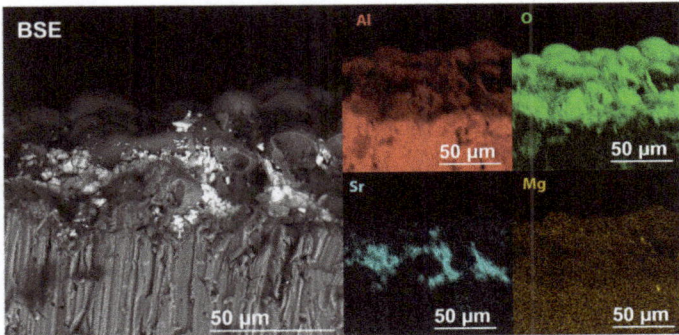

Figure 6. Cross-section BSE SEM image of the coating with element mapping for Al, O, Sr, and Mg.

3.2. Luminescence

To compare and study the optical properties of different samples, photoluminescence measurements were carried out in identical conditions for all the samples. To ensure that luminescence properties do not change after milling, the powder PL measurements were carried out for both powders. The PL spectra of the sample after pore filling were compared to powder samples and are displayed in Figure 7. All intensities have been normalized. Strontium aluminate powder before and after milling shows no significant difference in PL spectra, which means that the structure of the grains has not been disrupted; however, the absolute intensity decreased by 31%. The PEO sample with the 45-min treatment time after pore filling also exhibits the characteristic strontium aluminate luminescence band with a maximum at 530 nm. The additional peak can be seen at 693 nm; this includes the characteristic R1 and R2 lines of Cr^{3+} ion luminescence in an α-Al_2O_3 matrix (ruby). The intensity of this luminescence band correlates with the α-phase content, which in turn is related to the PEO processing time. The ion itself is acquired from the substrate, aluminium alloy, has the Cr as an additive in it. An increase in intensity in the near-IR region for the PEO sample is a second order of the UV part; the blue alumina luminescence is usually present in PEO coatings.

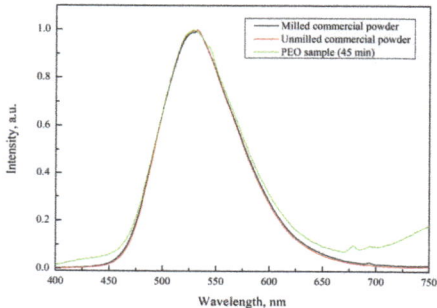

Figure 7. Photoluminescence spectra of powder before and after milling and 45-min PEO sample (P45Y).

In Figure 8, the PL spectra of samples with different PEO treatment times after pore filling are shown. The shape of the luminescence bands does not change except for the intensity of luminescence maximum. It can be seen that a longer PEO treatment duration leads to higher PL intensity, which is displayed in the insert of Figure 8. The PL intensity is directly related to the amount of strontium aluminate powder that can be incorporated into the pores and cavities of the coating. Samples with 15 and 20-min treatment time show almost identical PL intensity, while increasing the PEO processing time further increases the amount of powder that can be incorporated in the coating; a steady growth of luminescence is observed until 45 min of processing time (sample P45Y). One would expect that increasing the PEO treatment duration even further would lead to an increase in PL intensity; however, that is not the case. An even longer treatment time leads to the opposite effect, where the luminescence intensity drops significantly. This can be explained by the morphology of the coating; all the samples with a processing time longer than 45 min exhibit a transformation of the surface to a denser packed, less porous structure (can be seen earlier in SEM Figure 4f). The dense coating without cavities prevents the strontium aluminate powder adhering to the surface; there are no features for particles to attach to. To compensate for that, one can use smaller particles and fill the pores still present in a coating (e.g., black spots in Figure 4f); however, not only the overall intensity of the powder will decrease with the size of the particles, but the amount of powder that is possible to incorporate will decrease as well. This leads to the conclusion that 45 min at given PEO parameters is the optimal time for the chosen application. The 60-min sample (P60Y) is not presented in Figure 8 due to the low intensity of the signal, as no luminescent powder is present in the pores.

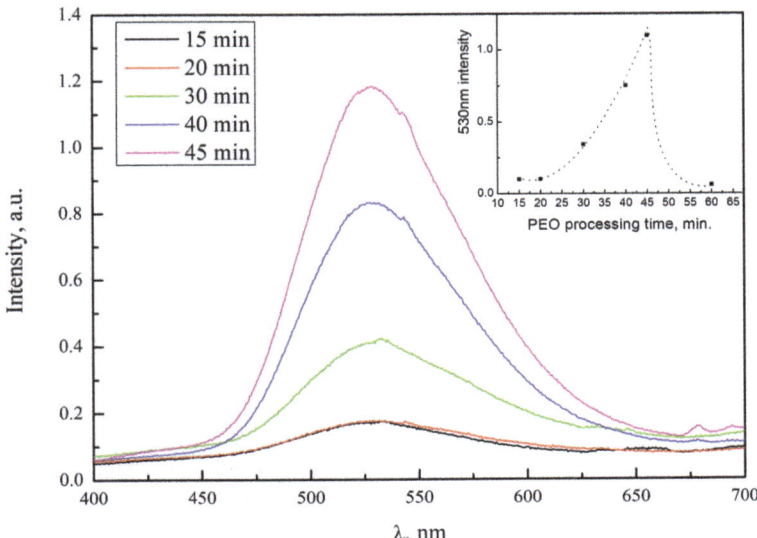

Figure 8. Photoluminescence spectra of samples after different PEO treatment durations. The inset shows the intensity dependence on the PEO processing time.

To perform luminescence decay kinetics measurements, each sample was irradiated with a 405-nm diode laser for 30 s. The kinetics measurements began 3 s after turning off the excitation source. Each measurement was carried out until the kinetics reached a plateau (the dynamic range of the detector is not enough to detect the changes in the intensity), which was around 4 min after ceasing excitation. Despite different initial intensities (shown in the inset of Figure 8), the luminescence decay kinetics characteristics for samples with various PEO treatment times did not vary (Figure 9), as the dual exponent approximation is essentially the same for all the samples. This leads to the conclusion that filling the powder into an aluminum oxide matrix does not disrupt the strontium aluminate powder structure, as expected.

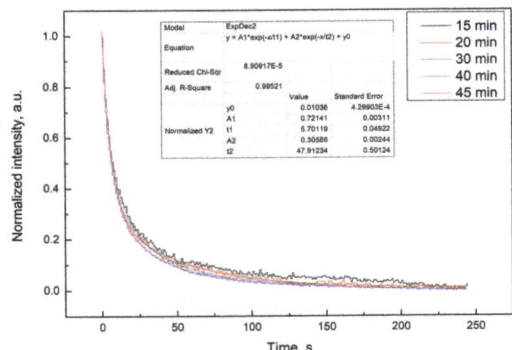

Figure 9. Decay kinetics measured for different samples. Dual exponent approximation results are presented in the inset table.

Although the PEO samples are visible in the dark for a few minutes, the duration of the afterglow is much shorter compared to commercial powder samples. This might be due to some surface effects that need to be studied in more detail.

4. Conclusions

During this study, it was shown that it is possible to produce phosphorescent PEO samples using activated commercial strontium aluminate powder and the pore-filling method. It was found that a longer PEO treatment duration leads to the development of larger and more densely packed pores and cavities, which in turn facilitates the filling process. The particle size of commercial strontium aluminate powder is too large to fit into the cavities of PEO coatings, so the powder needs to be milled to reduce the particle size. Luminescence measurements showed that milling and filling processes do not change the luminescence characteristics of the strontium aluminate powder, although the luminescence intensity is decreased. It was shown that PEO treatment duration influences luminescence intensity, which is directly related to the amount of powder that is incorporated into the coating. Treatment times until 45 min showed increasing luminescence intensity, but longer treatment times decreased luminescence intensity.

This method could prove useful for many practical applications where both surface protection and phosphorescent properties are necessary. Unfortunately, there are some drawbacks to this method. Although the substrate material is well protected from environmental effects, the incorporated powder is still vulnerable and prone to damage if exposed to water for prolonged periods of time. This means that the coatings should be additionally treated with some type of lacquer to reduce the degradation of the phosphorescent coating. Another obstacle that must be addressed in the future is how to ensure a larger amount of particles to incorporate into the coating and achieve more homogenous and denser distribution. Nevertheless, this method is promising and could be used as a substitute to currently used phosphorescent paints.

Author Contributions: Conceptualization, K.S. and D.M.; Data curation, K.A., A.Z., I.B. and K.L.; Formal analysis, I.B., K.L. and V.V.; Funding acquisition, K.S.; Investigation, A.Z., K.S. and D.M.; Methodology, K.A., A.Z., I.B., K.L. and V.V.; Project administration, K.S.; Resources, I.B. and K.S.; Software, A.Z.; Supervision, K.S.; Validation, K.A.; Visualization, K.A.; Writing – original draft, K.A. and K.S.; Writing – review & editing, K.A., K.S. and D.M.

Funding: This research project was supported financially by ERDF Project No: Nr.1.1.1.1/16/A/182.

Conflicts of Interest: The authors declare no conflict of interest.

References

1. Shionoya, S.; Yen, W.M. *Phosphor Handbook*, 1st ed.; CRC Press LLC: Boca Raton, FL, USA, 1999; pp. 655–658.
2. Lange, H. Luminescent europium activated strontium aluminate. US Patent US3294699A, 27 December 1966.
3. Matsuzawa, T. A New Long Phosphorescent Phosphor with High Brightness, $SrAl_2O_4:Eu^{2+},Dy^{3+}$. *J. Electrochem. Soc.* **1996**, *143*, 2670. [CrossRef]
4. Haranath, D.; Shanker, V.; Chander, H.; Sharma, P. Studies on the decay characteristics of strontium aluminate phosphor on thermal treatment. *Mater. Chem. Phys.* **2003**, *78*, 6–10. [CrossRef]
5. Lv, H.; Pan, Z.; Wang, Y. Synthesis and mechanoluminescent property of (Eu2+, Dy3+)-co-doped strontium aluminate phosphor by soft mechanochemistry-assisted solid-state method. *J. Lumin.* **2019**, *209*, 129–140. [CrossRef]
6. Peng, T.; Yang, H.; Pu, X.; Hu, B.; Jiang, Z.; Yan, C. Combustion synthesis and photoluminescence of SrAl2O4:Eu,Dy phosphor nanoparticles. *Mater. Lett.* **2004**, *58*, 352–356. [CrossRef]
7. Peng, T.; Huajun, L.; Yang, H.; Yan, C. Synthesis of SrAl2O4:Eu, Dy phosphor nanometer powders by sol–gel processes and its optical properties. *Mater. Chem. Phys.* **2004**, *85*, 68–72. [CrossRef]
8. Chang, C.; Yuan, Z.; Mao, D. Eu2+ activated long persistent strontium aluminate nano scaled phosphor prepared by precipitation method. *J. Alloys Compd.* **2006**, *415*, 220–224. [CrossRef]
9. Botterman, J.; Smet, P.F. Persistent phosphor SrAl2O4:Eu,Dy in outdoor conditions: saved by the trap distribution. *Opt. Express* **2015**, *23*, 868–881. [CrossRef]

10. Lu, X.; Blawert, C.; Kainer, K.U.; Zhang, T.; Wang, F.; Zheludkevich, M.L. Influence of particle additions on corrosion and wear resistance of plasma electrolytic oxidation coatings on Mg alloy. *Surf. Coatings Technol.* **2018**, *352*, 1–14. [CrossRef]
11. Sieber, M.; Simchen, F.; Morgenstern, R.; Scharf, I.; Lampke, T. Plasma Electrolytic Oxidation of High-Strength Aluminium Alloys—Substrate Effect on Wear and Corrosion Performance. *Metals (Basel).* **2018**, *8*, 356. [CrossRef]
12. Rokosz, K.; Hryniewicz, T.; Gaiaschi, S.; Chapon, P.; Raaen, S.; Matýsek, D.; Dudek, Ł.; Pietrzak, K. Novel Porous Phosphorus–Calcium–Magnesium Coatings on Titanium with Copper or Zinc Obtained by DC Plasma Electrolytic Oxidation: Fabrication and Characterization. *Materials (Basel)* **2018**, *11*, 1680. [CrossRef]
13. Apelfeld, A.V.; Ashmarin, A.A.; Borisov, A.M.; Vinogradov, A.V.; Savushkina, S.V.; Shmytkova, E.A. Formation of zirconia tetragonal phase by plasma electrolytic oxidation of zirconium alloy in electrolyte comprising additives of yttria nanopowder. *Surf. Coatings Technol.* **2017**, *328*, 513–517. [CrossRef]
14. Yerokhin, A.L.; Nie, X.; Leyland, A.; Matthews, A.; Dowey, S.J. Plasma electrolysis for surface engineering. *Surf. Coatings Technol.* **1999**, *122*, 73–93. [CrossRef]
15. Yerokhin, A.L.; Snizhko, L.O.; Gurevina, N.L.; Leyland, A.; Pilkington, A.; Matthews, A. Discharge characterization in plasma electrolytic oxidation of aluminium. *J. Phys. D. Appl. Phys.* **2003**, *36*, 2110–2120. [CrossRef]
16. Rahmati, M.; Raeissi, K.; Toroghinejad, M.R.; Hakimizad, A.; Santamaria, M. Effect of Pulse Current Mode on Microstructure, Composition and Corrosion Performance of the Coatings Produced by Plasma Electrolytic Oxidation on AZ31 Mg Alloy. *Coatings* **2019**, *9*, 688. [CrossRef]
17. Joost, W.J.; Krajewksi, P.E. Towards magnesium alloys for high-volume automotive applications. *Scr. Mater.* **2017**, *128*, 107–112. [CrossRef]
18. Huda, Z.; Taib, N.I.; Zaharinie, T. Characterization of 2024-T3: An aerospace aluminum alloy. *Mater. Chem. Phys.* **2009**, *113*, 515–517. [CrossRef]
19. Bouali, A.C.; Straumal, E.A.; Serdechnova, M.; Wieland, D.C.F.; Starykevich, M.; Blawert, C.; Hammel, J.U.; Lermontov, S.A.; Ferreira, M.G.S.; Zheludkevich, M.L. Layered double hydroxide based active corrosion protective sealing of plasma electrolytic oxidation/sol-gel composite coating on AA2024. *Appl. Surf. Sci.* **2019**, *494*, 829–840. [CrossRef]
20. Lu, X.; Chen, Y.; Blawert, C.; Li, Y.; Zhang, T.; Wang, F.; Kainer, K.U.; Zheludkevich, M. Influence of SiO2 Particles on the Corrosion and Wear Resistance of Plasma Electrolytic Oxidation-Coated AM50 Mg Alloy. *Coatings* **2018**, *8*, 306. [CrossRef]
21. Akatsu, T.; Yamada, Y.; Hoshikawa, Y.; Onoki, T.; Shinoda, Y.; Wakai, F. Multifunctional porous titanium oxide coating with apatite forming ability and photocatalytic activity on a titanium substrate formed by plasma electrolytic oxidation. *Mater. Sci. Eng. C* **2013**, *33*, 4871–4875. [CrossRef]
22. Grigorjeva, L.; Millers, D.; Smits, K.; Zolotarjovs, A. Gas sensitive luminescence of ZnO coatings obtained by plazma electrolytic oxidation. *Sensors Actuators A Phys.* **2015**, *234*, 290–293. [CrossRef]
23. Zolotarjovs, A.; Smits, K.; Laganovska, K.; Bite, I.; Grigorjeva, L.; Auzins, K.; Millers, D.; Skuja, L. Thermostimulated luminescence of plasma electrolytic oxidation coatings on 6082 aluminium surface. *Radiat. Meas.* **2019**, *124*, 29–34. [CrossRef]
24. Park, S.Y.; Jo, C.I.; Choe, H.-C.; Brantley, W.A. Hydroxyapatite deposition on micropore-formed Ti-Ta-Nb alloys by plasma electrolytic oxidation for dental applications. *Surf. Coat. Technol.* **2016**, *294*, 15–20. [CrossRef]
25. Kang, J.-I.; Son, M.-K.; Choe, H.-C.; Brantley, W.A. Bone-like apatite formation on manganese-hydroxyapatite coating formed on Ti-6Al-4V alloy by plasma electrolytic oxidation. *Thin Solid Films* **2016**, *620*, 126–131. [CrossRef]
26. Smits, K.; Millers, D.; Zolotarjovs, A.; Drunka, R.; Vanks, M. Luminescence of Eu ion in alumina prepared by plasma electrolytic oxidation. *Appl. Surf. Sci.* **2015**, *337*, 166–171. [CrossRef]
27. Stojadinović, S.; Tadić, N.; Vasilić, R. Photoluminescence of Sm2+/Sm3+ doped Al2O3 coatings formed by plasma electrolytic oxidation of aluminum. *J. Lumin.* **2017**, *192*, 110–116. [CrossRef]
28. Bite, I.; Krieke, G.; Zolotarjovs, A.; Laganovksa, K.; Liepina, V.; Smits, K.; Auzins, K.; Grigorjeva, L.; Millers, D.; Skuja, L. Novel method of phosphorescent strontium aluminate coating preparation on aluminum. *Mater. Des.* **2018**, *160*, 794–802. [CrossRef]

29. Martin, J.; Leone, P.; Nominé, A.; Veys-Renaux, D.; Henrion, G.; Belmonte, T. Influence of electrolyte ageing on the Plasma Electrolytic Oxidation of aluminium. *Surf. Coat. Technol.* **2015**, *269*, 36–46. [CrossRef]
30. Liepina, V.; Millers, D.; Smits, K. Tunneling luminescence in long lasting afterglow of SrAl2O4:Eu,Dy. *J. Lumin.* **2017**, *185*, 151–154. [CrossRef]
31. Aitasalo, T.; Dereń, P.; Hölsä, J.; Jungner, H.; Krupa, J.-C.; Lastusaari, M.; Legendziewicz, J.; Niittykoski, J.; Strek, W. Persistent luminescence phenomena in materials doped with rare earth ions. *J. Solid State Chem.* **2003**, *171*, 114–122. [CrossRef]

© 2019 by the authors. Licensee MDPI, Basel, Switzerland. This article is an open access article distributed under the terms and conditions of the Creative Commons Attribution (CC BY) license (http://creativecommons.org/licenses/by/4.0/).

Article

Fabrication and Characterization of Ceramic Coating on Al7075 Alloy by Plasma Electrolytic Oxidation in Molten Salt

Alexander Sobolev, Tamar Peretz and Konstantin Borodianskiy *

Department of Chemical Engineering, Ariel University, Ariel 40700, Israel; sobolev@ariel.ac.il (A.S.); tamarperetz15@gmail.com (T.P.)
* Correspondence: konstantinb@ariel.ac.il; Tel.: +972-3-9143085

Received: 6 September 2020; Accepted: 14 October 2020; Published: 17 October 2020

Abstract: The fabrication of a ceramic coating on the metallic substrate is usually applied to achieve the improved performance of the material. Plasma electrolytic oxidation (PEO) is one of the most promising methods to reach this performance, mostly wear and corrosion resistance. Traditional PEO is carried out in an aqueous electrolyte. However, the current work showed the fabrication and characterization of a ceramic coating using PEO in molten salt which was used to avoid disadvantages in system heating-up and the formation of undesired elements in the coating. Aluminum 7075 alloy was subjected to the surface treatment using PEO in molten nitrate salt. Various current frequencies were applied in the process. Coating investigations revealed its surface porous structure and the presence of two oxide layers, α-Al_2O_3 and γ-Al_2O_3. Microhardness measurements and chemical and phase examinations confirmed these results. Potentiodynamic polarization tests and electrochemical impedance spectroscopy revealed the greater corrosion resistance for the coated alloy. Moreover, the corrosion resistance was increased with the current frequency of the PEO process.

Keywords: plasma electrolytic oxidation (PEO); Al7075 alloy; aluminum oxide; molten salt; microhardness; corrosion resistance

1. Introduction

Modern industry widely applies the fabrication of different ceramic coatings on metallic substrates to obtain required properties. Usually, metals are implemented for wear and corrosion resistance. Nowadays, aluminum is the most promising metal since it is a main candidate to replace iron-based materials in various industrial applications. Among aluminum alloys, Al7075 alloy is a high strength alloy whose mechanical properties are comparable to many types of steel. Al7075 is applicable as aircraft fittings, shafts, and gears, valve components, and many other structural parts. However, this alloy has lower corrosion resistance than other aluminum alloys. This problem may be overcome by the development of a ceramic oxide coating on the Al substrate.

Plasma electrolytic oxidation (PEO) is one of the most promising environmentally friendly surface treatment processes to achieve the ceramic oxide coating on valve metals as Al, Mg, and Ti [1–6]. In PEO, a target metal is subjected to a high voltage which leads to the discharge appearance on the surface with the extremely high temperature and pressure that both provide oxidation of the surface. Usually, PEO treatment is conducted in an aqueous electrolyte made of silicates, phosphates, aluminates, fluorides, and other [7–12]. Recently, we have showed a possible implementation of the electrolyte of the molten salt in the PEO process [13–15]. Results of these works evaluated the formation of a denser coating free of any contaminants in comparison with a traditional method, which is usually obtained in an aqueous electrolyte.

Several works on the application of the PEO process on Al7075 alloy were recently published. Wang et al. investigated the corrosion resistance of scratched oxide and reported that scratches have reduced impedance and increased the corrosion current density [16]. Bahramian et al. reported the effect of TiO_2 nanoparticles addition to a PEO silicate-based electrolyte in Al7075 alloy [17]. The authors showed that the fabricated composite coating demonstrated improved corrosion resistance and mechanical properties due to the lower porosity content. Arunnellaiappan et al. investigated the effect of Al_2O_3 and ZrO_2 additions to an aqueous electrolyte on the coating formation. The authors showed improvement of a corrosion resistance of the fabricated coated alloy obtained in electrolyte with nanoparticles [18].

The aim of the current work is to fabricate and to characterize a newly formed oxide coating on Al7075 alloy using PEO in molten salt. The influence of the process current frequency on the structure and properties of the formed coatings was also analyzed. Morphology examinations and chemical composition were investigated by scanning electron microscopy (SEM) and energy dispersive spectroscopy (EDS). Phase composition was determined by X-ray diffraction (XRD) analysis. The corrosion resistance of the obtained coatings was studied by potentiodynamic polarization tests and electrochemical impedance spectroscopy (EIS).

2. Materials and Methods

Aluminum 7075-6 temper alloy (Scope Metal Group Ltd., Bnei Ayish, Israel) with chemical composition listed in Table 1 was used as the substrate for the PEO treatment. Investigated samples were of rectangle shape with 45×25 mm² and the thickness of 2 mm. Prior to PEO treatment, all specimens were ground using 400–1200 grit sandpapers, followed by ethanol cleaning and rinsing with distilled water.

Table 1. Chemical composition of the Al7075 alloy.

Chemical Element, wt.%							
Zn	Cu	Mg	Cr	Si	Mn	Ti	Al
5.1–6.1	1.2–2.0	2.1–2.9	0.18–0.35	<0.4	<0.3	<0.2	balance

The PEO treatment was carried out in a cylindrical furnace at a constant temperature of 280 °C in molten salt electrolyte with a eutectic composition of KNO_3–$NaNO_3$ (Sigma-Aldrich, St. Louis, MO, USA). The electrolyte was charged in a nickel crucible, which also acted as a counter electrode, while the Al alloy sample acted as a working electrode. The process was controlled by a MP2-AS 35 power supply (Magpulls, Sinzheim, Germany) with the electrical parameters: $I_{max} = 5$ A, $U_{max} = 1000$ V. Electrical parameters were pulsed at a frequency of 200 Hz (Sam-200 Hz), 300 Hz (Sam-300 Hz), and 400 Hz (Sam-400 Hz) with a duty cycle of 50% and recorded by Scope Meter 199C (Fluke, Everett, WA, USA). The process time was 30 min. After the treatment, all samples were washed with distilled water and dried in warm air.

The surface and the cross-section morphologies of the fabricated coatings were studied with SEM (MAIA3 TESCAN, Brno, Czech Republic). The elemental composition of the coatings was analyzed by EDS X-Max^N (Oxford Instruments plc, Abingdon, UK) detector in conjunction with the mentioned SEM. The phase analysis was determined by the X'Pert Pro diffractometer (PANalytical B.V., Almelo, the Netherlands) with Cu α radiation ($\lambda = 1.542$ Å) at the grazing incidence angle of 3° with a 2θ range from 20° to 90° (step size of 0.03°) at 40 kV and 40 mA.

The microhardness measurements were performed on cross-sections of the obtained coatings using a micro-hardness tester, Buehler Micrometer 2103 (Lake Bluff, IL, USA) under a load of 10 g. The average of five measurements for each oxide layer was presented in the results.

The corrosion resistance examination performed with a PARSTAT 4000A potentiostat (Princeton Applied Research, Oak Ridge, TN, USA). The potentiodynamic polarization test evaluated in 3.5 wt.%

NaCl (Sigma-Aldrich, St. Louis, MO, USA) solution at pH 7 using a three-electrode cell configuration wherein a Pt acted as a counter electrode and a saturated Ag/AgCl (Metrohm Autolab B.V., Utrecht, The Netherlands) acted as a reference electrode. The polarization resistance was detected at the range of ± 250 mV with the respect to the recorded corrosion potential at a scan rate of 0.1 mV/s. Prior to the test, all samples were kept in the solution of 3.5 wt.% NaCl for 30 min to reach the open-circuit potential (OCP) of a working electrode. The EIS measurements were performed at the OCP over a frequency range of 100 kHz to 1 mHz using a 5mV amplitude of sinusoidal voltage. The analysis of obtained spectra was made with the EC–Lab®software V11.10 fitting program.

3. Results and Discussion

3.1. PEO Processing

The PEO treatment is an electrochemical process which electrical parameters are shown in Figure 1. The electrical behavior shown on plots are almost the same for all three examined current densities. The received voltage, and the current, time behavior plots are typical plots for the unipolar PEO process. During the initial stage, for the first 50 s, the amorphous aluminum oxide coating was formed as the result of Al oxidation. This is clearly expressed on the plot where the current grew extremely. A further increase in the oxide coating led to the growth of the resistance in the substrate/electrolyte interface that resulted in current drop, as seen in the plot for 50–150 s. In the same period, the voltage increased up to 62 V. From this point of time (150 s), the formed amorphous coating transformed to crystalline, as expressed by a low current decrease and low voltage increase. The waveform of the process is expressed by a typical unipolar behavior where τ_{off} markup refers to the period when the current is not supplied and τ_{on} markup refers to the period when the current is supplied.

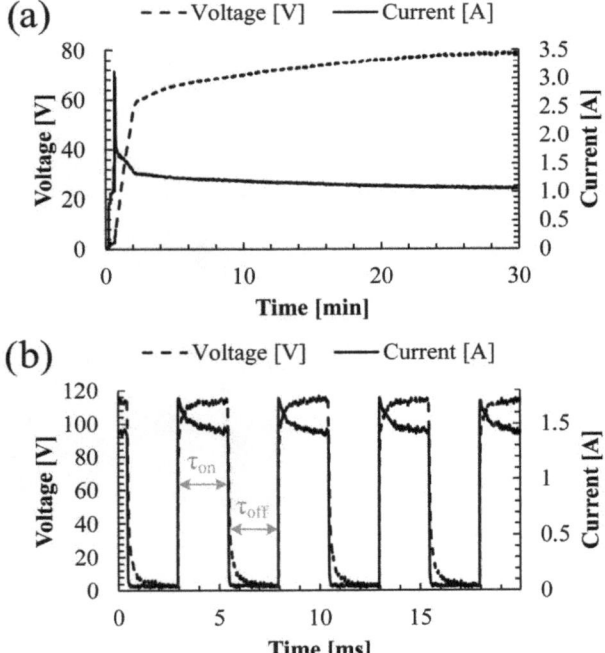

Figure 1. Plots of the PEO treatment of Al7075 alloy in molten salt: (a) voltage–and current-time behaviors, (b) waveform.

Surface morphology of the fabricated coatings for Sam-200 Hz, Sam-300 Hz and Sam-400 Hz are illustrated in Figure 2. Sam-200 Hz has a random-porous microstructure with sub-micron and micron pores. With the increase in the current frequency of the PEO, the microstructure changes, it contains only micron pores, which are more equal, and their size is larger. These changes are observed in a comparison of Sam-200 Hz with Sam-400 Hz microstructures (Figure 2a,c).

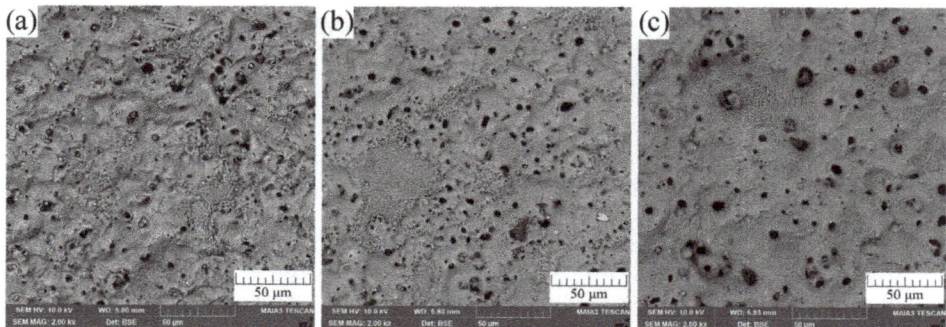

Figure 2. Surface morphologies of the Al7075 alloy after PEO treatment obtained by SEM: (**a**) Sam-200 Hz, (**b**) Sam-300 Hz, (**c**) Sam-400 Hz.

Almost the same thickness of the coating, around 25 μm, was revealed in cross-section images in Figure 3. Evaluation of these images revealed a double-layered coating structure which consists of the internal dense layer and the external porous layer. These layers are attributed to two different phases of aluminum oxide, α-Al_2O_3 and γ-Al_2O_3 that to be discussed in Section 3.3. Observation of these images also revealed that the coating of the Sam-400 Hz is denser and less porous, which is correlated to the morphology observation discussed before. These structural changes may be attributed to the energy impact of PEO treatment. With the current frequency increase, the time duration of each single current pulse decreases that leads to the reduction in the transferred energy. Hence, the volume of the locally re-melted coating decreased, and thereby it became denser.

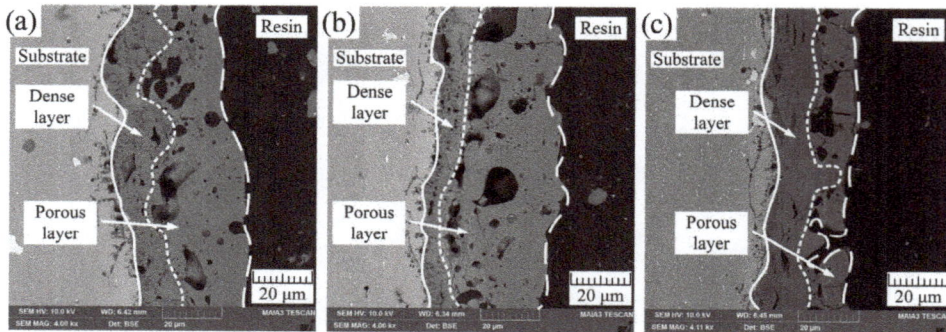

Figure 3. Cross-section images of the Al7075 alloy after PEO treatment obtained by SEM: (**a**) Sam-200 Hz, (**b**) Sam-300 Hz, (**c**) Sam-400 Hz.

3.2. Phase and Chemical Composition Characterization

XRD analysis was performed to evaluate phase composition of the fabricated coatings; its patterns are shown in Figure 4. The following phases were detected in three examined coatings: α-Al_2O_3, γ-Al_2O_3, and metallic aluminum which also corresponds with [19,20]. Minor changes in peak intensities were observed due to the different absorbance of the X-ray radiation that may be attributed to a non-uniform fabricated surface. As expected, two phases of aluminum oxide were detected, and they are referred to as the internal and the external layers of the coating. Metallic aluminum phase was identified from the substrate due to the high penetration depth of the X-rays. Furthermore, no other phases were found, which points to the formation of a pure oxide coating with no impurities which usually originates from the electrolyte decomposition.

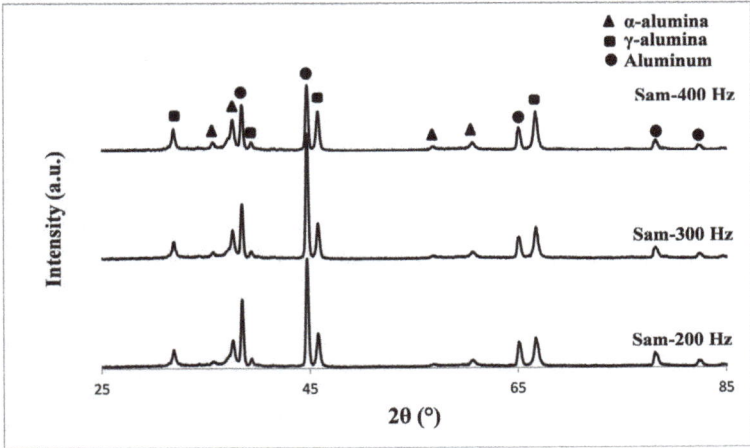

Figure 4. XRD patterns of the Al7075 alloy surfaces after PEO treatment of Sam–200 Hz, Sam–300 Hz and Sam–400 Hz.

EDS analysis evaluated presence of mostly aluminum and oxygen elements and tiny amount of magnesium and zinc as shown in Figure 5. As expected, Al and O are the main components of the Al_2O_3 coating, Mg and Zn are components originated from the Al7075 alloy substrate. EDS mapping images pointed on the high O content which is favorably located in the coating (dark red color in the coating). Observation of a metallic substrate revealed presence of mostly Al, Zn, and Mg elements (darker colors for these three elements in the substrate). These observations are well correlated with the XRD patterns and point to the formation of the oxide coating free of any contaminates. These undesired components are usually originated in the decomposition of an aqueous electrolyte.

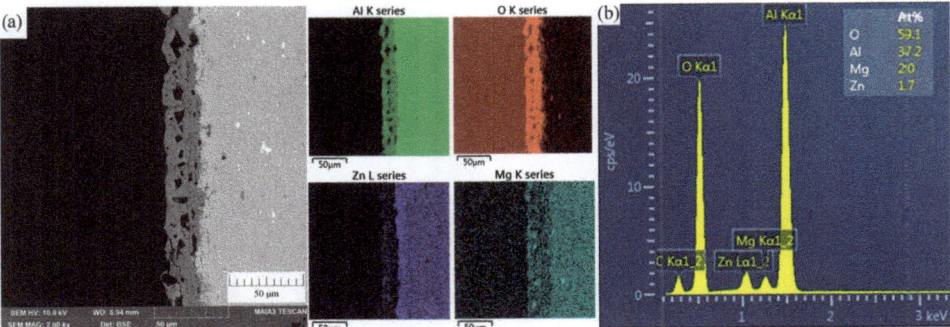

Figure 5. SEM image of the Sam-400 Hz after PEO treatment: (**a**) cross-section and EDS elements mapping, (**b**) EDS analysis.

3.3. Coating Characterization

Microhardness measurements results are illustrated in Figure 6. These results exhibit increase of the coating microhardness with the current frequency increase. Additionally, the external layers show higher values in comparison with the internal. The measured values for the internal layer were 473.08, 490.30, and 588.67 HV for Sam-200 Hz, Sam-300 Hz, and Sam-400 Hz, respectively. The measured values for the external layer were 847.60, 967.77 and 1112.93 HV for Sam-200 Hz, Sam-300 Hz and Sam-400 Hz, respectively. The microhardness measurements jointly with the XRD results may indicate locations of the formed oxide layers. As reported by Brabec et al. and Sobolev et al. [15,21], the external layer is referred to as the α-Al_2O_3 since it is harder than the γ-Al_2O_3.

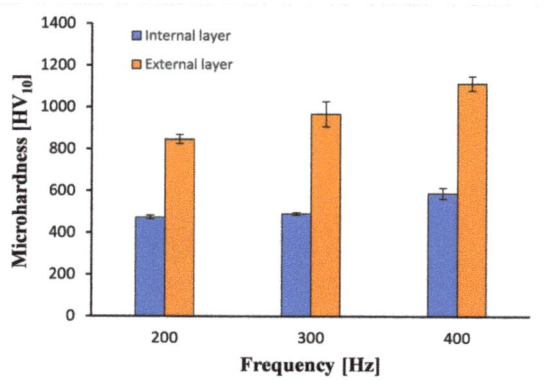

Figure 6. Microhardness measurements of internal and external layers of the Al7075 alloy coatings after PEO treatment of Sam-200 Hz, Sam-300 Hz and Sam-400 Hz.

The fabricated oxide coating provides a high corrosion protection [22–24]. Two methods of corrosion resistance evaluation were conducted, the potentiodynamic polarization method and the EIS.

Potentiodynamic polarization curves for three coated samples and the original alloy are illustrated in Figure 7 and the obtained values are listed in Tables 2 and 3.

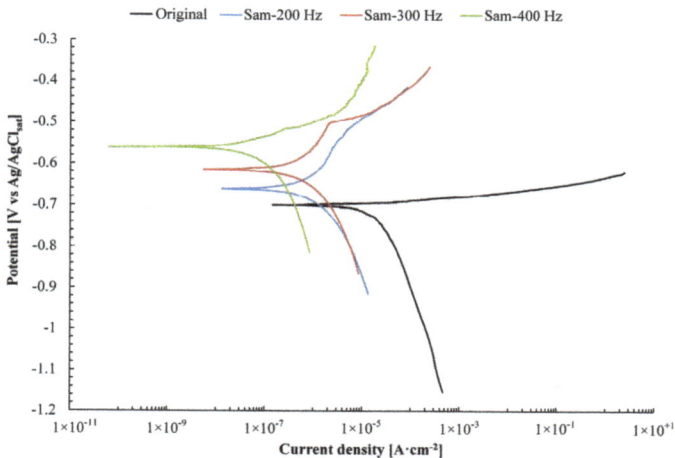

Figure 7. Potentiodynamic polarization curves of Sam-200 Hz, Sam-300 Hz, Sam-400 Hz and an original Al7075 alloy. Examination was conducted in the 3.5 wt.% NaCl solution.

Table 2. Mean of electrochemical parameters obtained from the potentiodynamic polarization curves for the fabricated and original alloys.

Samples	$i_{corr} \times 10^{-7}$ (A·cm^{-2})	E_{corr} vs. Ag/AgCl$_{sat}$ (mV)	β_a (mV/dec)	β_c (mV/dec)
Original	248.5	−701.2	17	−324
Sam-200 Hz	8.9	−654.7	199	−164
Sam-300 Hz	7.6	−605.9	207	−222
Sam-400 Hz	2.3	−560.6	76	−457

Table 3. Calculated values of the resistance and the corrosion rate from the potentiodynamic polarization test for the fabricated and original alloys.

Samples	R_p (kΩ cm^2)	CR (mm/Year)
Original	1.5	0.3192
Sam-200 Hz	43.7	0.0092
Sam-300 Hz	61.4	0.0083
Sam-400 Hz	122.9	0.0026

Potentiodynamic results evaluated that all coated alloys exhibit greater corrosion resistance than the original alloy. Thus, the polarization curves were shifted to lower current densities as observed in Figure 7. Moreover, evaluation of the curves revealed that the current density shift was increased with the increase in current frequency of PEO process. In other words, the oxide surface can effectively prevent the penetration of corrosive solution. The obtained values for the corrosion potential (i_{corr}) presented in Table 2 were determined from the Tafel plots in Figure 7, while other values in Table 3 were determined from the Stern–Geary equation:

$$R_p = \frac{\beta_a \times \beta_c}{2.3 \times i_{corr}(\beta_a + \beta_c)} \quad (1)$$

where R_p refers to the polarization resistance, β_a and β_c refer to the anodic and cathodic Tafel slopes, respectively, and i_{corr} refers to the corrosion current density.

The corrosion rate (CR in mm/year) was calculated from the following equation:

$$CR = \frac{k_r \times i_{corr} \times EW}{\rho \times A} \quad (2)$$

where k_r the corrosion rate constant (3272 mm/(A·cm·year)), i_{corr} is the corrosion current in amperes, EW is the equivalent weight in g/equivalent, ρ is density in g/cm^3, and A is the area of the sample in cm^2.

The corrosion resistance values in Table 3 revealed the greater corrosion resistance for treated alloys using PEO in molten salt. Additionally, the effect of the PEO current frequency on the corrosion resistance was also revealed. It was determined that the increase in current frequency caused to the increase in corrosion resistance; the current density reduced from 8.9×10^7 A·cm^2 for Sam-200 Hz to 2.3×10^7 A·cm^{-2} for Sam-400 Hz. The same tendency was also found for the corrosion potential which was increased from −654.7 to −560.6 mV, while the original alloy corrosion potential was −701.2 mV. The calculated corrosion resistance rates of the coated alloys are higher than the rate of the original alloy. Hence, the calculated corrosion rate of the original alloy was 0.3192 mm/year while the coated alloys exhibited corrosion rates of 0.0092, 0.0083, and 0.0026 for Sam-200 Hz, Sam-300 Hz, and Sam-400 Hz, respectively. Experimental results of the present work have also revealed that the obtained corrosion resistance in PEO conducted in molten salt is much higher than the identical alloy treated using PEO in aqueous solution [25]. This behavior is attributed to the surface morphology where substrate/coating interface characterized by the electric double layer with a specific ion adsorption.

The corrosion process in a chloride-based solution was extensively shown elsewhere [26–29]. These works described the activation effect of ions of Cl$^-$ in the anodic dissolution of aluminum and aluminum oxide. Lv et al. [30] and Zhang et al. [31] reported the following corrosion reactions which can be also adopted to the present investigation:

The anodic reaction:

$$Al \rightarrow Al^{3+} + 3e^- \tag{3}$$

with the possible subsequent reactions

$$Al^{3+} + 3H_2O \rightarrow Al(OH)_3 + 3H^+ \tag{4}$$

$$Al(OH)_3 + Cl^- \rightarrow Al(OH)_2Cl + OH^- \tag{5}$$

$$Al(OH)_2Cl + Cl^- \rightarrow Al(OH)Cl_2 + OH^- \tag{6}$$

$$Al(OH)Cl_2 + Cl^- \rightarrow AlCl_3 + OH^- \tag{7}$$

The cathodic oxygen depolarization reaction:

$$O_2 + 4e^- + 2H_2O \rightarrow 4OH^- \tag{8}$$

It may be assumed that the cathodic slope of the Tafel plot indicates the diffusion control for the reaction of the oxygen reduction (reaction 8). Meanwhile, the anodic slope of the Tafel plot indicates the activation-controlled process of the charge transfer through the interphase. Based on the theory, a one-electron process is the most preferable, as shown in reactions (5)–(7). It is worth to add that the anodic dissolution of the original alloy occurs uniformly, without pitting corrosion. However, dissolution of the coated alloys occurs with the local pitting corrosion. Different slopes of the βa for the coated alloys are attributed to the different overpotentials obtained at a certain current decade and they determine the corrosion rate of the process (Equation (2)).

EIS analysis revealed the effect of the PEO current frequency on the surface morphology formation. EIS presents Nyquist plots in Figure 8, Bode plots in Figure 9, and equivalent electrical circuits in Figure 10 and Table 4 which provide the fitting between the experimental and theoretical results.

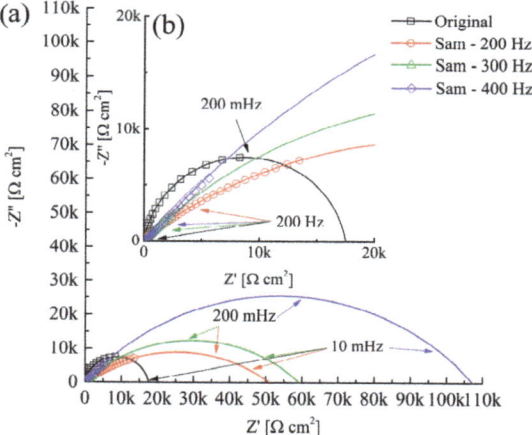

Figure 8. Nyquist plots of Sam-00 Hz, Sam-300 Hz, Sam-400 Hz and an original Al7075 alloy: (**a**) curves for full impedance range, (**b**) enlarged area of the curves. The symbols represent experimental values and the solid lines represent fitted data. Examination was conducted in the 3.5 wt.% NaCl solution.

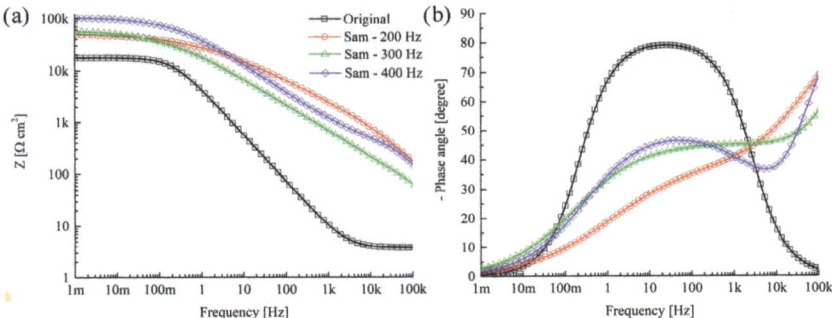

Figure 9. Bode plots of Sam–200 Hz, Sam–300 Hz, Sam–400 Hz and an original Al7075 alloy: (**a**) impedance modulus, (**b**) phase angle. The symbols represent the experimental values and the solid lines represent fitted data. Examination was conducted in the 3.5 wt.% NaCl solution.

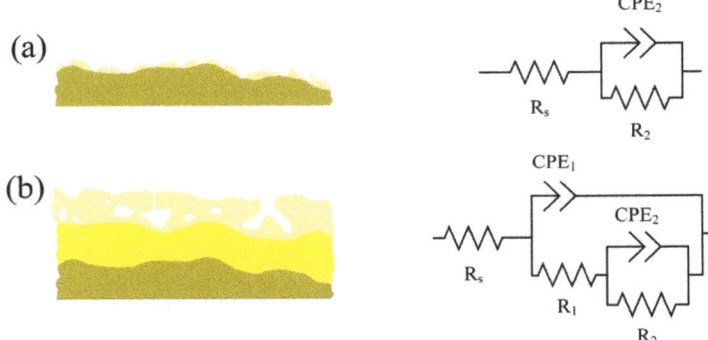

Figure 10. Equivalent electrical circuits used for the EIS spectra fitting for: (**a**) original Al7075 alloy, (**b**) Al alloys treated by PEO in molten salt.

Table 4. Fitting values of electrical parameters obtained from the equivalent electrical circuits for the fabricated and original alloys.

Samples	Original	Sam-200 Hz	Sam-300 Hz	Sam-400 Hz
R_s ($\Omega \cdot cm^2$)	3.72	0.56	0.33	0.79
CPE_1 ($F \cdot cm^{-2} \cdot s^{n-1}$)	-	1.35×10^6	1.55×10^{-6}	1.86×10^6
n_1	-	0.63	0.61	0.63
R_1 ($\Omega \cdot cm^2$)	-	1.21×10^2	1.52×10^2	2.56×10^2
CPE_2 ($F \cdot cm^{-2} \cdot s^{n-1}$)	4.27×10^5	5.84×10^7	1.05×10^8	1.62×10^8
n_2	0.90	0.88	0.61	0.57
R_2 ($\Omega \cdot cm^2$)	1.75×10^4	5.11×10^4	5.30×10^4	1.07×10^5
χ^2	1.01×10^3	1.12×10^3	1.42×10^3	1.25×10^3

The Nyquist plots illustrated in Figure 8 represent differences between behavior of the coated samples and the original alloy. The symbols in the plot are attributed to the experimental values while the solid lines are the fitted data presented in Table 4. The Nyquist plot for the original alloy has a one capacitive loop which is referred to as the natural aluminum oxide layer. However, treated alloys using PEO have two capacitive loops, the first one is located at the high-frequency range and referred to the external outer layer. The second capacitive loop is located at the medium and low-frequency ranges, and referred to as the internal oxide layer.

Fitting of the EIS spectra was made by equivalent electrical circuits given in Figure 10. Here, R is the charge transfer resistance and CPE is the constant phase element of the electric double-layer capacitance determined using the follows equation:

$$Z_{CPE} = \frac{1}{(Qj\omega)^n} \tag{9}$$

where Z_{CPE} refers to the impedance of the CPE, Q refers to the proportional factor of the CPE, j is the imaginary unit, ω is the angular frequency, and n is a dimensionless parameter. When $n = 0$, the CPE acts as a resistor. When $n = 1$, the CPE acts as a capacitor.

The investigation of the natural aluminum oxide layer on the original alloy was performed by the equivalent circuit $Rs + CPE_2/R_2$ which is illustrated in Figure 10a. Rs is the resistance of the electrolyte, R_2 is the charge transfer resistance, and CPE_2 is the electric double-layer capacitance. The investigation of the treated alloys using PEO was performed by the equivalent circuit $Rs + CPE_1/(R_1 + CPE_2/R_2)$ which illustrated in Figure 10b. R_1 is the resistance to the charge transfer of the external porous layer, CPE_1 is the electric double-layer capacitance of the porous external layer, R_2 is the resistance to the charge transfer of the internal dense layer, and CPE_2 is the electric double-layer capacitance of the internal dense layer.

An evaluation of Bode plots in Figure 9 revealed presence of two bends on the curves of the treated alloys that may be attributed to the internal- external- layered oxide structure. Similar to the description of capacitive loops in Nyquist plots, in Bode plots, the curve at the high- frequency range is referred to as the external outer layer and that at the medium- and low-frequency ranges are referred to as the internal layer.

An electrical parameters analysis in Table 4 revealed that values of R_2 are much greater in comparison with the values of R_1 that may be attributed to the greater corrosion resistance, preferable provide by the internal layer. For Sample-400 Hz, R_2 reaches 107 kΩ·cm^2 while the value of the R_1 for the same sample is 256 Ω·cm^2. The same trend was also observed for Sam-200 Hz and Sam-300 Hz. Thus, Sam-200 Hz exhibits values of 51.1 kΩ·cm^2 and 121 Ω·cm^2 for R_2 and R_1, respectively. Sam-300 Hz exhibits values of 53.0 kΩ·cm^2 and 152 Ω·cm^2 for R_2 and R_1, respectively. These results point to the high corrosion-resistance performance of the internal dense oxide layer which is located closer to the substrate and acts as a barrier for the electrolyte penetration towards the metallic substrate. Values for the external layer are referred to as the porous surface morphology and correspond with observations in Figure 3. Sam-200 Hz and Sam-300 Hz demonstrate almost the same porosity while Sam-400 Hz has lower porosity as determined in SEM images and fitted values of the electrical parameters.

The calculated values for n varied from 0.57 to 0.88 for the treated alloys due to the non-uniform coating structure. The value of n for the original alloy is 0.9, closer to the capacitor behavior.

4. Conclusions

This work focuses on an investigation of the PEO process carried out in molten salt. Here, the ceramic coating on Al7075 alloy was fabricated and characterized. SEM observation revealed a porous surface which changes from the random sub-micron and micron porosity to a more equal micron porosity with the increase in current frequency of the process. The thickness of the coatings was around 25 µm and it contained two oxide layers. Based on XRD and EDS analysis and microhardness measurements, the external layer was detected as α-Al$_2$O$_3$ and the internal as γ-Al$_2$O$_3$. The corrosion resistance of the original and coated alloys was examined by a potentiodynamic polarization approach and EIS. Both methods revealed the greater corrosion resistance for the coated alloy than the original one. Additionally, it was determined that the PEO current frequency affects morphology, and as a result, its corrosion resistance. Thus, the higher the current frequency of the PEO process, the greater the corrosion resistance of the alloy.

Author Contributions: Conceptualization, A.S. and K.B.; methodology, A.S. and K.B.; investigation, A.S and T.P.; writing—original draft preparation, A.S., T.P., and K.B.; writing—revised draft preparation, A.S. and K.B.; supervision, K.B. All authors have read and agreed to the published version of the manuscript.

Funding: This research received no external funding.

Acknowledgments: Authors wish to express their gratitude to Alexey Kossenko and Natalia Litvak from the Engineering and Technology Unit at the Ariel University for their assistance in XRD and SEM investigations.

Conflicts of Interest: The authors declare no conflict of interest.

References

1. Stojadinović, S.; Vasilić, R.; Petković, M.; Kasalica, B.; Belča, I.; Zekić, A.; Zeković, L. Characterization of the plasma electrolytic oxidation of titanium in sodium metasilicate. *Appl. Surf. Sci.* **2013**, *265*, 226–233. [CrossRef]
2. Lu, X.; Blawert, C.; Kainer, K.U.; Zheludkevich, M.L. Investigation of the formation mechanisms of plasma electrolytic oxidation coatings on Mg alloy AM50 using particles. *Elect. Acta* **2016**, *196*, 680–691. [CrossRef]
3. Dehnavi, V.; Liu, X.Y.; Luan, B.L.; Shoesmith, D.W.; Rohani, S. Phase transformation in plasma electrolytic oxidation coatings on 6061 aluminum alloy. *Surf. Coat. Technol.* **2014**, *251*, 106–114. [CrossRef]
4. Rokosz, K.; Hryniewicz, T.; Matýsek, D.; Raaen, S.; Valíček, J.; Dudek, L.; Harničárová, M. SEM, EDS and XPS analysis of the coatings obtained on titanium after plasma electrolytic oxidation in electrolytes containing copper nitrate. *Materials* **2016**, *9*, 318. [CrossRef]
5. Yerokhin, A.L.; Nie, X.; Leyland, A.; Matthews, A. Characterization of oxide films produced by plasma electrolytic oxidation of a Ti–6Al–4V alloy. *Surf. Coat. Technol.* **2000**, *130*, 195–206. [CrossRef]
6. Park, S.Y.; Choe, H.C. Functional element coatings on Ti-alloys for biomaterials by plasma electrolytic oxidation. *Thin Solid Films* **2020**, *699*, 137896. [CrossRef]
7. Gnedenkov, S.V.; Khrisanfova, O.A.; Zavidnaya, A.G.; Sinebrukhov, S.L.; Gordienko, P.S.; Iwatsubo, S.; Matsui, A. Composition and adhesion of protective coatings on aluminum. *Surf. Coat. Technol.* **2001**, *145*, 146–151. [CrossRef]
8. Asquith, D.; Yerokhin, A.; James, N.; Yates, J.; Matthews, A. Evaluation of residual stress development at the interface of plasma electrolytically oxidized and cold-worked aluminum. *Metall. Mater. Trans. A* **2013**, *44*, 4461–4465. [CrossRef]
9. Simchen, F.; Sieber, M.; Kopp, A.; Lampke, T. Introduction to plasma electrolytic oxidation—An overview of the process and applications. *Coatings* **2020**, *10*, 628. [CrossRef]
10. Lou, H.-R.; Tsai, D.-S.; Chou, C.-C. Correlation between defect density and corrosion parameter of electrochemically oxidized aluminum. *Coatings* **2019**, *10*, 20. [CrossRef]
11. Yang, K.; Zeng, J.; Huang, H.; Chen, J.; Cao, B. A Novel self-adaptive control method for plasma electrolytic oxidation processing of aluminum alloys. *Materials* **2019**, *12*, 2744. [CrossRef]
12. Terleeva, O.P.; Slonova, A.I.; Rogov, A.B.; Matthews, A.; Yerokhin, A. Wear resistant coatings with a high friction coefficient produced by plasma electrolytic oxidation of al alloys in electrolytes with basalt mineral powder additions. *Materials* **2019**, *12*, 2738. [CrossRef]
13. Sobolev, A.; Kossenko, A.; Borodianskiy, K. Study of the effect of current pulse frequency on Ti–6Al–4V alloy coating formation by micro arc oxidation. *Materials* **2019**, *12*, 3983. [CrossRef] [PubMed]
14. Sobolev, A.; Kossenko, A.; Zinigrad, M.; Borodianskiy, K. Comparison of plasma electrolytic oxidation coatings on Al alloy created in aqueous solution and molten salt electrolytes. *Surf. Coat. Technol.* **2018**, *344*, 590–595. [CrossRef]
15. Sobolev, A.; Kossenko, A.; Zinigrad, M.; Borodianskiy, K. An investigation of oxide coating synthesized on an aluminum alloy by plasma electrolytic oxidation in molten salt. *Appl. Sci.* **2017**, *7*, 889. [CrossRef]
16. Wang, S.; Gu, Y.; Geng, Y.; Liang, J.; Zhao, J.; Kang, J. Investigating local corrosion behavior and mechanism of MAO coated 7075 aluminum alloy. *J. Alloys Compd.* **2020**, *826*, 153976. [CrossRef]
17. Bahramian, A.; Raeissi, K.; Hakimizad, A. An investigation of the characteristics of Al_2O_3/TiO_2 PEO nanocomposite coating. *Appl. Surf. Sci.* **2015**, *351*, 13–26. [CrossRef]
18. Arunnellaiappan, T.; Arun, S.; Hariprasad, S.; Gowtham, S.; Ravisankar, B.; Rama Krishna, L.; Rameshbabu, N. Fabrication of corrosion resistant hydrophobic ceramic nanocomposite coatings on PEO treated AA7075. *Ceram. Int.* **2018**, *44*, 874–884. [CrossRef]

19. Nie, X.; Meletis, E.; Jiang, J.; Leyland, A.; Yerokhin, A.; Matthews, A. Abrasive wear/corrosion properties and TEM analysis of Al_2O_3 coatings fabricated using plasma electrolysis. *Surf. Coat. Technol.* **2002**, *149*, 245–251. [CrossRef]
20. Polat, A.; Makaraci, M.; Usta, M. Influence of sodium silicate concentration on structural and tribologicalproperties of microarc oxidation coatings on 2017A aluminum alloy substrate. *J. Alloys Compd.* **2010**, *504*, 519–526. [CrossRef]
21. Brabec, L.; Bohac, P.; Stranyanek, M.; Ctvrtlik, R.; Kocirik, M. Hardness and elastic modulus of silicalite-1 crystal twins. *Micropor. Mesopor. Mat.* **2006**, *94*, 226–233. [CrossRef]
22. Wen, L.; Wang, Y.; Zhou, Y.; Ouyang, J.-H.; Guo, L.; Jia, D. Corrosion evaluation of microarc oxidation coatings formed on 2024 aluminium alloy. *Corros. Sci.* **2010**, *52*, 2687–2696. [CrossRef]
23. Shen, D.; Li, G.; Guo, C.; Zou, J.; Cai, J.; He, D.; Ma, H.; Liu, F. Microstructure and corrosion behavior of micro-arc oxidation coating on 6061 aluminum alloy pre-treated by high-temperature oxidation. *Appl. Surf. Sci.* **2013**, *287*, 451–456. [CrossRef]
24. Venugopal, A.; Srinath, J.; Rama Krishna, L.; Ramesh Narayanan, P.; Sharma, S.C.; Venkitakrishnan, P.V. Corrosion and nanomechanical behaviors of plasma electrolytic oxidation coated AA7020-T6 aluminum alloy. *Mater. Sci. Eng. A* **2016**, *660*, 39–46. [CrossRef]
25. Rao, Y.; Wang, Q.; Oka, D.; Ramachandran, C.S. On the PEO treatment of cold sprayed 7075 aluminum alloy and its effects on mechanical, corrosion and dry sliding wear performances thereof. *Surf. Coat. Technol.* **2020**, *383*, 125271. [CrossRef]
26. Evans, U.R. CXL—The passivity of metals. Part, I. The isolation of the protective film. *J. Chem. Soc.* **1927**, 1020–1040. [CrossRef]
27. Rosenfeld, I.L.; Marshakov, I.K. Mechanism of crevice corrosion. *Corrosion* **1964**, *20*, 115t–125t. [CrossRef]
28. Hoar, T.P.; Mears, D.C.; Rothwell, G.P. The relationships between anodic passivity, brightening and pitting. *Corros. Sci.* **1965**, *5*, 279–289. [CrossRef]
29. McCafferty, E. General relations regarding graph theory and the passivity of binary alloys. *J. Electrochem.* **2003**, *150*, B238–B247. [CrossRef]
30. Lv, D.; Ou, J.; Xue, M.; Wang, F. Stability and corrosion resistance of superhydrophobic surface onoxidized aluminum in NaCl aqueous solution. *Appl. Surf. Sci.* **2015**, *333*, 163–169. [CrossRef]
31. Zhang, B.; Wang, J.; Wu, B.; Guo, X.W.; Wang, Y.J.; Chen, D.; Zhang, Y.C.; Du, K.; Oguzie, E.E.; Ma, X.L. Unmasking chloride attack on the passive film of metals. *Nat. Commun.* **2018**, *9*, 2559. [CrossRef] [PubMed]

Publisher's Note: MDPI stays neutral with regard to jurisdictional claims in published maps and institutional affiliations.

© 2020 by the authors. Licensee MDPI, Basel, Switzerland. This article is an open access article distributed under the terms and conditions of the Creative Commons Attribution (CC BY) license (http://creativecommons.org/licenses/by/4.0/).

MDPI
St. Alban-Anlage 66
4052 Basel
Switzerland
Tel. +41 61 683 77 34
Fax +41 61 302 89 18
www.mdpi.com

Coatings Editorial Office
E-mail: coatings@mdpi.com
www.mdpi.com/journal/coatings